Global City Review

To Make a Long Story Short

FALL 1994 · ISSUE NUMBER FOUR

GLOBAL CITY REVIEW
To Make a Long Story Short
FALL 1994
NUMBER FOUR

Global City Review is published twice yearly.
SUBSCRIPTION PRICE: $12 FOR ONE YEAR, $20 FOR TWO YEARS.
INTERNATIONAL: $17 FOR ONE YEAR, $30 FOR TWO YEARS.
INSTITUTIONS: $15 FOR ONE YEAR, $25 FOR TWO YEARS.
MAKE CHECKS PAYABLE TO GLOBAL CITY REVIEW

ALL CORRESPONDENCE SHOULD BE SENT TO:
GLOBAL CITY REVIEW
SIMON H. RIFKIND CENTER FOR THE HUMANITIES
THE CITY COLLEGE OF NEW YORK
138TH AND CONVENT AVENUE
NEW YORK, NY 10031
PLEASE INCLUDE A SASE WITH ALL MANUSCRIPTS.

GLOBAL CITY REVIEW IS DISTRIBUTED BY
BERNHARD DEBOUR, INC.

ISSN. 1068-0568

WITH ONGOING THANKS TO
THE SIMON H. RIFKIND CENTER FOR THE HUMANITIES
AT THE CITY COLLEGE OF NEW YORK,
AND TO ANN EUGENIA VOLKES.

DESIGN BY CHARLES NIX

Contents

ESSAYS

AN INTERVIEW

A MEMOIR

POEMS

CONTENTS

To Make a Long Story Short:

ELEVEN QUESTIONS

What does it mean to make a bouillon cube of art?

Where does one art end and the next begin?

Can the writer put all of life into one sentence?

What is the meaning of blind?

Are male and female distinct?

What happens to a word in isolation?

Are children strangers to their parents?

What do we forget when we put pen to paper?

Can parents plot their children's lives?

Where is the poetry in untimely death?

If war wins, where is art?

E.M.BRONER, HEATHER RAMSDELL, *and*
ANNE VATERLAUS, *Issue Editors*

Before

by Khaled Mattawa

SOMEWHERE BETWEEN FAITH AND GRACE THERE is always the footprint of logic lost in the purest light. Not hidden at all, but a vehicle, a necessity, neither mop, nor bucket, but whatever gives the floor its shine. Repeat after me: The sun through the window pours on the floor, and the wood glistens as if in praise. As if a child breaking into a run. That is what I see through the window now. A child breaking into a run for the simple flame that must burn and because there are such words as "snuffed out." I could be weeping and wailing because the child is not a memory, only a gesture on my part. Yesterday, I fed a friend's cat and talked to her because the town has been emptied and filled with snow embroidered with tire tracks. I fed a friend's cat and she rubbed her sides against my calves. The thing to say now is that I am in the middle of a life in a house with the owners on holiday. Or to say a car engine hums (the owner forgetting the keys inside), and is on its way to a crystalline loss. Here deduction is howling at an on-coming storm. The thing is I fed a friend's cat and later poured a bowl of milk for her and she sniffed it, barely licked it, and left. The thought is. The life is. I've visited graves—tombstones 10 feet high. I ran through the cemetery and laughed my Cairo laugh. I wanted to be arrested by the police, wanted someone to take down what I had to say. Whatever I would have said would have been the truth. But there was no one there. Only dusk and a shitload of romance. Only dusk and the dry tongue of the interstate. Detroit, Toledo, and the befores that follow the first before. Billions of snowflakes in between. I slam the brakes. I veer. The radio plays a Nigerian song. I feed the cat and talk to her and I take the milk away and begin to forget. This was before. The first before. Days later I dissolve in the fumes of Houston. There are no snowflakes in Texas and Muzak fills the emptiness in between. Your father died last week I said. Oh cat why do you stare at me and why do I stare back. Emptiness stares at the cat and the cat stares at the missing bowl and he stares at the after beyond the first before. Repeat after me: A Nigerian song fills your town, fills the betweens with a billion befores.

3 poems from *The Cry of Merlin*

by Paul Pines

Neruda's voice
rises from the page
grieving for

Lorca, Alberti and Mayakovsky;

likens them
to a river, a mountain and a crystal,
and himself

to a valley
with subterranean waters
pooling at its heart.

It is night.

I ride through the dark
on a bus
past little boxes
of light,

in awe
of this solitude

I call
my life.

memory is
ADJECTIVE
TO TIME

I have walked a Mayan city
in the jungle
near Benque Viejo—

a syntax
of chambers
walls and patios
under half an inch of earth

buried in my mind
like a message
waiting

to be smuggled back into the world

Orpheus
loses Eurydice
to return
                    with the vision
                    of Eurydice
which dissolves like a word
on his tongue

So he goes down again to
the forge
                    where forms
                    are made
to emerge
with a song that contains

a moth
playing at the source
of desire—
                        a river of fire
that flows both ways

Yugoslavia

by Eliot Weinberger

WHEN YOU OPEN THE BOOK, THE BOOK EXPLODES in your face. A severed head on a pile of shoes.

War war and always wars, but there are certain wars that seem evidence of something more than the varieties of human brutality. Certain small wars, seem from a distance that turns their daily horror into an allegory: another, bloodless war that is reenacted in the mind, our minds, we who are not in the slaughter. Wars one can't stop thinking about. As if the way they, over there, are dying is a reflection of the way we are living. And worse: that the way we are living is the cause of their dying.

Until now, Spain, in this century, was the small internal war that unravelled transnationally as an allegory of the age. Not only as the first performance, in a provincial "theater of war," of the global cataclysmic battle against fascism: Spain, in Western minds, was the triumph of the destroyers of art, and of the technologically powerful over the aspirations of the ordinary people. "Guernica," the painting, draws its force, not from its depiction of carnage—there is always carnage—but from the image of the old ways being obliterated by what is most horrible in the new ways. The airplane—till then the great symbol and glory of the Machine Age—brings not the reign of progress, but a rain of bombs. It is a scene that was reenacted from the conquest of the

Americas to the last of the colonial wars: Algeria, Vietnam, Angola. What we, the "modern" people, most hate about ourselves is the agent of the destruction of what we imagine is most admirable in the "old" people, the people we think we once were.

These are wars that seem, for the moment, over: the imperial territorial expansions, the wars in the name of this century's barbaric ideologies, the aspirations of a peasantry hopelessly crushed by the technologically advanced. Instead we have the so-called "tribal" wars: the revivals, manipulated by the power-hungry, of supposedly ancient ethnic, racial, religious animosities. Wars, we are told, that return to the roots of war.

But these are not wars taking place in the rain forest, or even in the Balkans in the years before the First World War. These are wars being acted out within a global network of communication, and within a global consumer market. The dead are seen "live," all over the planet, on the television news; the combatants, on either side, drink the same Coca-Cola. The world is simultaneously coming together and blowing apart.

Yugoslavia—ex-Yugoslavia—above all, is the emblem of the age, and not merely as the chaos that follows the collapse of empires and ideologies. Yugoslavia is the nervous breakdown of the collective mind of the West.

Sick and weary at the end of a century that murdered millions in the name of certainty, the generation that considers itself "post"—post-modern, post-ideology—has, in an unprecedented, nearly unbelievable manner, transformed doubt into a science. After three centuries, the Age of Criticism has reached its decadence. Words do not mean what they say: books are lies imagined into truth by their readers: images are the representation of a conspiracy between creators and receptors; every narrative is false, and any opposing notion equally true. It is a dismantling of the world, they say, so that murderous certainty will never happen again.

As always, what will never happen happens over. The tragedy of Yugoslavia is the certainty of the combatants, and the paralyzing uncertainty of the rest who are watching. There, people die every day in the name of belief, and continue to die every day in the name of disbelief. Here, we see the evidence, and wonder what is evidence, and wonder what to do, and wonder in what name is anything done, and let a thousand other images rush in.

"Happy"

by Joe Elliot

Happy

to think I'm not real where
else could it come from

“the flower”

by Joe Elliot

the flower

does nothing

so well

then fails

“If words were ever definite enough”

by Joe Elliot

If words were ever definite enough
to refer only and exactly to what
the speaker is thinking of
then the lover having glimpsed would slowly be wrapped
round and round in the endlessly long
letter he must write until immobile on his back
a pharaoh in a cenotaph

Recording

by Sari Broner

admitting to herself that she didn't know   realizing it was a new sound
one she hadn't heard the elephants make before   remembering something else
she'd felt throbbing between the sounds

she went back   to pick up everything the equipment could   recording
the baby elephants at high speed   slowing it down   in the sound room   able
then to hear the deep rumblings of the baby elephants   to hear other
things she now realized were not the elephants   things she couldn't
identify   that were calling

Seven Out of Ten

from Number One Son, *a novel*

by Shan Wu

SEAN
Morning
Hollywood, California

I am not as gifted as you were. Even if I had not taken my long hiatus from the Art, I could not be you.

"Seven out of ten," Uncle Randy finally said when I pushed him once to rate me as a martial artist. "Seven out of ten."

When I came back to the Art at age 17, I went to Uncle Randy Ng's school in Seattle. It was a night where the rain fell in hard little streams. The industrial grey streets around the school were deserted, the pavement felt slippery and soft. I'd walked all the way across town from Queen Anne with no umbrella. I wore your leather jacket—the one decorated with your ink drawings of men in martial arts poses. It was the last year I could wear any of your clothes—I was outgrowing them fast. The school sat at an intersection of streets filled by boxy buildings with flat roofs. It rose into my sight as I started down a hill on the street opposite the intersection. A small white blob, the male half of the *yin-yang* symbol, stood out dully against the dark of the building. Fluorescent light reached silently out of the second floor windows like beckoning fingers.

"Hearing is more important than sight in fighting," you used to say.

This made sense to me for the first time as I gazed at the school through the rain. The sounds I associated with training—breath blowing through lungs, the rustling of loose clothes, grunts and impact slaps—all these were absent. In their place was only the constant rattle of water against pavement, walls, roofs, leaves, branches, my hair and your jacket. Enveloped in that sound the school seemed small and fake. Cardboard scenery on a too bright stage. You needed sound to understand what sight told you. I wished my sister, Sarah, were there so I could talk to her about this.

Sarah doesn't remember as much as I do which frustrates her. I tell her how lucky we are that you always spoke to us as though we could understand everything. Even if we didn't understand then, it becomes clear one day. I tell Sarah to be patient, that she'll remember more later.

I huddled beneath the door frame, my back pressed up against the cold black metal door, facing the street. Above me the *yin-yang* symbol floated like a halo. I felt the rivets of the door prod me gently through the wet leather jacket. I worried about the ink running. Sarah, I was thinking, got the athletic genes. At 14, she was a state champion gymnast and an aspiring dancer. Strength and grace. At 17, I was thin and moody. A muffled impact slap came through the door—it sounded like somebody kicking a heavy canvas bag. The old sound reverberated through me, warming my back and hands. I remembered a patch of green grass and sunlight tickling my bare shoulders as I mimicked your stance. Your hands were outstretched, palms facing inward as though holding up small mirrors for you to admire yourself in. I opened the metal door and went into the narrow hallway preceding the steep wooden stairs.

The sounds were all there. The ceiling above me creaked with human effort.

Going Down Slow

by Cornelius Eady

MY DADDY IS REFUSING TO EAT. HE HAS BEEN IN THE hospital too long; he has had to watch as one of his many roommates, a quiet black gentleman about his age, passes away in the night, he has to lie alone in a bed and watch his own torso dwindle away; to be patient as his pants and shoes become obsolete.

As soon as he eats, he has to go to the bathroom, and the man can no longer walk, cannot rise, an adult, from his bed; cannot shut the bathroom door behind him, the word *private*, impossible magic, never to be his again.

The man my daddy has become cannot deal with more than a thimble of food, and he will not get adjusted to a world of harassed nurses and public bedpans. He will let most of his meals lie fallow on their trays.

He will glare at his son's persistence with the anger the sick have for what a healthy body can't know.

He will force you to smell the blossoming of his terrible resolve.

He would rather die.

Money (that's what I want)

by Cornelius Eady

MY FATHER HAS JUST DIED, AND NOW I MUST stand in line at the Manhattan branch of his bank in order to withdraw funds from his account before they find out and freeze them in probate.

My daddy is gone, but if I want to give him a proper burial, I must begin this last dance with him. I must stuff my pockets with his mojo, the greenback dollar, the one true spell he ever believed in, and then walk two awkward blocks from his bank to mine.

The teller takes his time checking me out. He can smell that something isn't quite right, but since my power of attorney is in working order we both know how this transaction must end.

He must count my father's leavings into my hand. How transparent my pockets will feel under the midday New York sun. How nervous my steps as I carry him.

A Field

by Deirdre Kovac

I AM THINKING OF YOU FROM THE PASSENGER SEAT in a car on the highway to New York. The Pennsylvania countryside passes at a steady rate, catches in the corner of my window—the horizon a clothesline gripped at both ends and gently shaken at two different tempos: the curve of my eye, the speed of the earth. The road rises and a field unfolds as we near the top of a hill.

An ice-blue house in the middle distance is surrounded by short grass. Irrigation ditches point to the stoop, resembling perspective. Or Bruegel without revelers. I think of Ohio and from there to you with several steps along the way fallen between the cracks. It isn't seamless, a train ride that skirts the drab industrial shore of Lake Erie. And even then the station houses are small and pristine as if I couldn't have looked out the window the whole way between them. Not that I think of you each time I pass this field.

The line that leads me from this field to you cannot be called a thread. A thread does not crack and if it is torn we see the strands were meant to overlap. Memory moves forward over a wooden floor just as the boards are each laid down and toed in place leaving gaps, giving them room to move when the weather changes.

It is early winter here and the road is a dry, medium grey. The color of cement. The basement floor of your house. As if colors and days followed in a sequence of precise gradation from red to violet—not this drastic leap off the highway, through a field, to a roundhouse outside Kent where I picture you on the railroad tracks, walking and then waiting. Now it is so quiet I hear the car engine, a muted holler crowding the forest with brisk empty shapes. They stay the same if we would stand still.

Discrete bits of snow cover the grass in a series of unnamed constellations. They are out of view before the dots can be connected. The road descends and closes the field. The house disappears. Did I tell you it was blue? Grey grows more vibrant the farther I get from the thing itself.

Curly Travel

by Karen Garthe

Tennessee
Miles in air
Miles cut rock
For miles trees turned to blue
Their orange speeding in
Where light goes and
Come-on night, You
Parallel Gorge
Of empty laps
Classic pitch
The driver hums
Poor loves the moon rounds
Red-eyed
Passing
His laurel unlikely
Hood of rain

Carpenter

by Karen Garthe

Like cradles,   like roaring
on the birthday floor
barges above fields *of wine   like cradles*
grave scrapes & guilds

Of The Carpenter who builds loud
mystery clocks
who's sad
effigies scare
described Innocence

The ark's messy
jigsaw
chairs in stump loneliness
skeins
of repetition blanks in the coda
It's Revery
like cradles
the lamp to your pages
Fumble in the glass

Legacy

by Ann T. Rosato

1961

THERE'S SOMETHING ABOUT THAT HOUSE, THE one in the middle of the block with the wooden porch and blinds that are always closed. Neighborhood children run by the house and even loiter in front of the black iron gate, but they are never invited in. Grass is not allowed to grow. Instead, the front yard is made of thousands of small stones, some round, some rough and jagged. A narrow cement path leads to the front porch; it's bordered by grey molding to prevent any wild thing from poking through.

A man lives there alone with his children. He has three with long yellow braids and pale palms. Their eyes are difficult to see because they are always cast down; their feet are never free of shoes and socks and their bodies are dressed carefully: plaid skirts and slips, even in summer. The three girls hold hands as they walk two blocks to school, silent and slow, father watching them from the porch. They do not skip, or break out into a run; their walk is steady.

The oldest girl has many friends inside of her and they comfort and protect her. They take the blame when that is required, shield her from small spaces, broom handles and sweaty legs wrapped around her. She jumps at shadows and sometimes cannot remember going to school or waking up in the morning. Her grey eyes flicker and blink quickly when something is wrong, or out of place. She has the face of her mother the day she left them, the face of a stone goddess: beautiful, empty and determined never to return. She holds her sisters' hands tightly, their sweaty fingers sticking to hers, their nails digging into the backs of her own white hands, but still, she will not let them go. The oldest is a hawk watching over her young, hovering always.

1972

My mama left us alone in the Palms Motel, just next to the freeway, in Downey, California; room number 16. At night I could hear the engines of cars and trucks racing along the freeway, and the voices of the women in the courtyard outside. The women's noises were soft, but there was something nasty in them as they offered men beers and the feel of their nightgowns. The engines were more comforting.

I told mama not to leave us, but she patted my brown hair and pointed to the Winchell's bag of doughnuts on the edge of the sofa bed.

"I'll be back by supper; we'll eat at McDonalds, O.K.?" She was asking for my permission.

"O.K., but you'll be back before dark?"

"Of course, honey...watch your sister!"

My sister was six months old and I did watch her, all the time. Who knew when she'd swallow some tiny thing that would stick in her throat, or roll off the bed or something. I couldn't lose her, I knew that.

Mama straightened her belt, reminded me of the diapers in the corner and slipped quietly out the door. Mama always came and went silently; I couldn't even hear her high heels clicking on the walk out front. I put the chain on the door and began to clean up by putting everything in the middle of the card table. I paid special attention to the small objects that littered the floor: bottle caps, lighters, burnt matches. My baby sister was sleeping on the twin bed we shared and I checked her every five minutes to be sure she was still breathing. She was so quiet. Mama had always warned me that babies could die "just like that!" and she'd snap her fingers in my face.

The police found us when the manager had gotten tired of waiting to be paid. He was a large Indian man who wore white Polo shirts and shorts all the time. I wondered if he owned anything else or if his closet was full of white shirts and shorts, clean and crisp like new. He found us asleep on the bed when he let himself in. We had been alone for two days with a dozen Winchell's doughnuts, a jar of peanut butter and half a can of Enfamil. I had cut the doughnuts into small pieces, but they were still almost gone. The manager said we smelled bad and the police officers put on rubber gloves before they touched us.

I was five years old that year and I didn't see my mother again 'til 13 years later. 1985, on my eighteenth birthday, like a wish or a curse, she appeared out of nowhere and congratulated me. But I had grown to 18 without her and smiled, said thank you and waved to her from a car full of girlfriends, candy bars and Prince tapes.

1961

The closets in the house all have locks on them, to keep possessions inside, and the bathroom doors all have large round holes where the doorknobs should be, because little girls shouldn't have secrets. The bedrooms have no doors and though each girl has her own bed, they sleep together in their thin nightgowns, their socks curled around each other.

The house is quiet, really too quiet for a house with children. They play tea parties and cut out Debbie Reynolds paper dolls. They giggle and whisper in each other's ears and read the Bobsey Twins out loud to one another. But they never know if they've been bad, so they always stay quiet, giggle behind their hands and read under the covers.

The middle girl is invisible. She pulls her knees tightly against her chest, clasps her hands flat against them and tucks her head into her body. This is how she sleeps: invisible. She flattens herself against walls, slithers noiselessly through doorways and pushes herself into the recesses of the sofa, losing herself deep into the cushions. When someone calls her, she does not answer immediately, but waits to see if they will call her once and forget her, or if they will persist. She dreams of foreign countries and languages she cannot understand.

1988

My baby sister has always been irresponsible. She was 14 when she first got pregnant. I won't let a man touch me. She says I won't let anyone touch me. She had an abortion, and then got pregnant again a year later. She was five months pregnant before she realized it. I said maybe she lets everyone touch her. She tells me she's cold at night, and afraid; a blanket isn't enough. On her 16th birthday, I give her a down comforter, though now she has a baby to keep her warm.

I go to college and live with my cousin. Her mother, my aunt, is staying at a residential treatment facility. That's a vague term for a looney bin. We both know what it is, and that they won't ever let her out, but we don't talk about that. My cousin visits her mother every Saturday. I refuse. My aunt gives me the creeps; the way her grey eyes blink and change suddenly. I have to study anyway. It's safe in the library or the lab. I pour things, burn things, look up formulas: there's an order to everything I do. As long as I can read, I don't have to stay where I am. I never read stories about motels.

1961

The inside of the house is always dark and the heat is never turned on. In the winter, they shiver and hold each other, in the summer they just shiver. It seems it is always cold. Inside the closet doors, hanging on nails, are three belts, each one displaying a name on the smooth underside. The names are painted with red nail polish and the letters meticulously spell out each girl's name. Identical sets hang in each closet so they won't forget.

They set the table for dinner each night and they make their own beds each morning. Everything is ordered and scheduled. They have learned how to do things properly, and on time, but still there are sometimes mistakes. The rings in the bathtub are hard to clean completely, and the ammonia stings their fingers.

The youngest one is the witness. She looks least like her mother, and most like her father. She has fleshy cheeks and large blue eyes that seem to glisten, as if always wet. Her bottom lip trembles so slightly, she's told it's charming. She is the guilty one and she does not have nightmares. There is nothing more that scares her. She is the reason the others are punished, she is a lion in a steel cage: powerful, but trapped. She watches, her own white skin never bruised, never scratched. There are no burn marks on her feet, or scars on her belly. There is only clean, fair skin, soft yellow hair and responsibility.

1992

My baby sister finally entered into a drug treatment program. Our young aunt, who calls herself a missionary, drove her to the place in her red Volkswagen, chirping to her the whole way about leading a responsible life. The place is an old stucco building with fluorescent lights and cafeteria trays. My sister's kid is with our grandfather in the same house our mother and aunts grew up in. It's pretty run down now; there are no doors on the bedrooms, and it always seems cold in there. The kid is almost four and asks too many questions. It's not even the questions that really bother me, it's that he doesn't ever want to hear the answers. My grandfather is an old man with baggy pants and slippers that seem to be glued to the bottoms of his feet. I don't see the kid much, he doesn't ever remember my name. I'm about to start a Master's program at USC and I need to concentrate on that anyway.

I thought I saw my mother the other day; the lady was pushing a grocery cart full of aluminum cans and used cardboard pizza boxes. I went up close, and she smiled at me. I gave her a dollar when I realized it wasn't her. Two days later I got a small brown envelope in the mail. It had been forwarded several times and the thin manila paper had tears in it where pens had crossed out old addresses. Inside was a letter telling me my mother is dead. The things inside the envelope were the things they found in her pockets. There's a small black key, a blank postcard of Victoria Falls in Zimbabwe and a tiny square photo like you get in a photo booth. At first I don't recognize myself. Then I see my baby sister, six months old in black and white, her hand stuffed in her mouth, her eyes wide from the flash. She is sitting on my mother's lap, while my mother poses. I am standing to the side, squeezing my body into the square piece of glass that's the frame. We all look tired, but we're smiling. I wonder where the other three pictures are, they usually come in strips of four.

Magic Carpet Edition

from The Book of Knowledge

by Barbara Louise Ungar

WONDER QUESTIONS:
1) Does light die away? 2) Does the moon
pull the sea? 3) What makes us sneeze?
4) Who invented the zipper? 5) Where do all the stones
come from? 6) Why does furniture make noise
at night? 7) What *is* linoleum?
8) What was wrong with Achilles'
heel? 9) Is it darkest
just before dawn? 10) Is it true that sound
goes on forever? 11) Will the world
become like the moon? 12) Why do our voices sound
hollow in an empty hall?

ANSWERS:

12) To answer this
question, we might begin by asking ourselves why we use
the word "hollow" to describe our voices. 8) Nothing. 3)
It's a reflex. 11) Our earth will probably, in time,
become a dead world in the heavens,
with not a breath of air. 1) Light always dies
away when its source ceases, yet light
might travel on forever, spreading like ripples in a pond. 2) Yes.
6) It's always groaning; we only listen at night
when we can't sleep. 9) No. 5) Stones are
pieces of broken rock. 7) Linseed oil & cork.
4) Whitcomb L. Judson. 10) No sound
lasts forever *as a sound*. Yet nothing
is ever lost: the energy of its waves must go on in some way forever.

The Trip of the Column Through the Wall

by Diane Lewis

> *"Gravity is to Architecture what Mythology is to Literature."*
>
> —E.A. POE

1. An iconography of a succession of masculine and feminine gods was nuanced through the elements of plinth, column, and wall, accumulating a lesson of position and relation inspired by each memorable source loci given by earth to man's conscience. A freestanding poem to place composed of autonomous elements exposed in light against the image of earth's dramatis personae.

2. A necklace of sacred places around the rim of a sea is constructed by locating pragmatic bases for fueling stops as the interstitial itinerary. The specificity of place retains a sacral standing, in contrast to systemic structure relegated to the pragmatic. Temples of a previous legacy compress into perimeter gates; the column not only stands in front of the wall, it enters the wall and reappears imbedded as the pilaster. The elements form identities positioned in dialogue within the walls.

3. Baptism is necessary to reinvigorate empire. The font is placed in the basilica. The internal colonnade stands witness

within walls. The ribcage externally expresses the inner light.

4. A new conception of volume encloses all historic layers positing porous boundaries. Internal colonnade pivots at points of intersection. Christian acknowledges pagan. Sacraments substitute for triglyphs and trophys. Pagan logic brings freedom to faith. The vertical axis remains vacant to force a centripetal layering of alternating boundaries and shells of structure. Walls and columns rhythmically succeed.

5. Walls break free of correspondence to columns. Skin is autonomous from bones. Diverse principles exist in simultaneity. Exposed skeleton and skin are possessed by light vectors. Visions of illumination result.

6. Contradictory orders exist in parallel and at varied scales. Elements are mapped at multiple scales and addressed to different proximities. An infinite field is achieved; boundaries become spheres of influence. A hieroglyphic expressive of the mythology of existence rotates up from plan. Gravity reveals its face.

The Sighting

by Ann Lauinger

I almost missed the great turtle's quiet
shimmy off the bank and into the pond:
mud, carapace, and water for an instant
three incarnations of one substance, three
siblings or three images from an alchemist's dream
  the chocolate earth
    the bottle-brown-green translucency
      the gold-flecked disc of tortoise shell.

On stumpy flippers it ferried itself out
and stopped midway between the shores,
floating ponderously; two tiny bulges
on the end of a chewed cigar were watchful.
Shocking, the improbable grace of an ugly thing.
  And then it turned
    its back on me, heaved up its rear
      and plummeted with the certainty of stone.

That day I thought the turtle was my emblem,
voyager between worlds, no more at home
in light and air than lapped like a rock by the glistening
dappled water or inching along the murk.
I longed to dive with it down to what I fear—
  mud in the nostrils

    between the clenched teeth, mud
        mud in the throat, my beginning—

to struggle back to the surface spitting out
tadpoles sweetgrass pearls prayers my mother's names.
What I need to know is in the slime
is at the bottom where the turtle goes
is in the broth it paddles in and feeds on.
    It eats and holds
        the mystery digested and dispersed:
            it is the elements it moves among.

Today I think, my angel not my emblem.
Only a god can know and be at once,
untroubled as a stone. From a life of slow
incessant journeying the turtle retrieves
nothing, knows nothing, has nothing to show
    except survival
        and that beautiful petrifaction
            token of long practice in the art of descent.

Prehistory

by Ann Lauinger

He waits for his son. Behind walls of glass,
on every side a draped or looping snake.
Though some are small and not one is awake,
his hackles rise at the sheer collective mass
of muscle. Pythons, thick as a man's thigh
and coiled in careless figure-eights like rope
of nylon green, repeat the ancient trope
of evil: self on self returning.

Am I

not free? Silence. His coward heart ricochets,
a frantic bird assailing its cage of bone—
when sudden laughter twists him round to face
fat snakes festooning outstretched arms. *Make known*
your name, exultant stranger! Let me praise
you, joy and terror triple-twined, my own.

Some Minor Notes

by Louise Corbett

FIONA WAS NOT HER MOTHER'S BEST PUPIL, NOT by any means, but she had certain talents, her mother said, certain strengths: a light touch, a good ear. Easily bored, Fiona even played her scales dramatically. Fiona herself hated the piano though drawn to it; it was a way of expressing her unhappy spirits: minor notes, minor chords.

"I don't want to play in the recital this year," said twelve-year-old Fiona; she had just finished a two-piano duet with her mother, Fiona playing the melodic line, her mother, the lower heavier chords requiring a wider reach than Fiona could manage.

"Why ever not?" inquired her mother, turning an astonished face toward Fiona.

"I don't want to, that's all." She was too big, too old. At her mother's side she found herself changing into an apprentice servant, passing the cookie plate, pouring out the fruit punch, keeping an eye on the younger children. "And anyway, I'm not that good," she added.

"You could be good," her mother countered. "If you practiced more, you'd be good," which Fiona had heard before.

But the next day, she came home from school to find her mother sitting on the piano bench crying. Simultaneously horrified and embarassed, she stood dumbly by. Finally she put a hand on her shoulder. What could the matter be?

"I want you to play in the recital this year. I want you to play your little Allegro." So Fiona played the Haydn. It was another two years before her mother stopped requiring her to play in the year-end recital.

2

Fiona's mother came out on the front porch a little after ten on an early spring evening. The porch light was on and she was in her nightclothes, her blue robe sashed around her waist, her greying hair, unpinned and hanging down her back. She was not crying so much, a few sighs, a few gasps, but the tears were starting to flow and they were followed by deep, gut-wrenching sobs. Pacing and turning, she moved back and forth from one end of the porch to the other, pivoting by the front door. She wiped her eyes on the sleeve of her robe every now and then but most of the time her hands were clasped together, sometimes she actually seemed to be wringing them. Meanwhile, she talked steadily if disjointedly, odd phrases, muffled by the sobs but strangely magnified at this hour of the night. By half-past the hour, she was crying luxuriantly, passionately, as the furies cry, denied their due.

A car door slammed and Fiona's broad bland face emerged from the half-darkness of the city street. She was almost fourteen, pale and thin; her thick fair hair was elaborately and unbecomingly curled and her mouth was covered with several layers of dark red lipstick. She wore a sweater and skirt, there were nine or ten bangle bracelets on her thin wrists and her nails were dirty with traces of polish around the edges. She had gone to a school party, she was only asked to dance twice and she was late because the father of a friend drove her home and he didn't arrive at school until half an hour after the dance was over. The apparition on the front porch may have been no more than a shadow to him, but it was not one he was anxious to meet, and he drove rapidly off. The girl cast one all-encompassing look at her mother shuffling pathetically towards her, the tangled hair, the gaping mouth, the eyes, swollen with tears, the bathrobe with its trailing sash, and, pausing no longer than a heartbeat, pushed open the door, entered the house and rapidly climbed the stairs. Once in her room she closed the door, threw her clothes onto a chair, and went immediately to bed, lights out, eyes firmly closed. The sooner she slept, the sooner the day would be over.

Her mother, meanwhile, moving more slowly, closed and locked the front door and turned off the porch light. Then she went into the kitchen, filled a teacup with sherry, and carried if off with her to bed.

3

It was Fiona's first year of college and she was home for Christmas. She had not changed very much, she was a little taller, a little heavier. The bangle bracelets were gone, however, and so was the nail polish. She had stopped curling her hair and it fell to her shoulders unadorned. She used rather a lot of lipstick still, and a little mascara as well. She sat with her mother at the breakfast table and talked about her school work, her courses: economics, chemistry. Not doing so well in economics, not doing so well in chemistry.

"Well, what are you doing well in, my God?"

"Oh, well," said Fiona, drinking her tea. "My social life is picking up anyway," she added, moving smoothly into the one subject that currently did interest her.

"You always had plenty of social life."

"Did I? You think?" Fiona was neutral on the subject. She took another biscuit, slathered on the butter, ladled on the honey.

"I should know if anyone should," said her mother.

Crunch, crunch went the biscuits, delicious honey-saturated crumbs. Taking her time, Fiona chewed and swallowed, cleared the crumbs from her throat. "OK, well now it's better."

"Is it," said her mother, neutral in her turn.

"I have this new boyfriend," Fiona said, getting right to it.

"What about Phil?"

"What about him?"

"I thought you were writing to Phil?"

"Why would I write to Phil?"

"Well, the way you acted, the way you carried on..."

Fiona moved on. "So anyway, I'm going to..." (down went the last mouthful of biscuit, crunch, crunch, crunch) "visit his family..." (crunch, crunch) "the last part of vacation..."

"When?" Her mother produced this with an incredulous stare; she may well have been reviewing the various methods she used in the past to control this wayward daughter. The stare was part of her repertoire; until fairly recently, Fiona was unable to meet her stare head on.

"The last four days, I forget. I have to look it up, I think it's the twenty-ninth?"

"I will absolutely not allow it."

Now Fiona was drinking her tea, swilling it down; not

ordinarily a noisy eater, she was even making slurping noises.

"You eat like this at school I suppose?"

"In Maine," Fiona said. "He lives in Maine. So it's not too much out of the way," she added thickly through a mouthful of biscuit.

"If your father were alive, he would be far stricter than I am, you know that," her mother announced.

Fiona received this remark in silence. If true, she was certainly glad he was dead. She got up and began clearing away the dishes. Subject closed, she seemed to say.

As far as her mother was concerned, all Fiona did during this vacation was read novels and talk on the telephone. Occasionally she appeared to be reading her history book. Or she sat on the floor playing solitaire.

Her mother stood looking down on her. "You never even go near the piano," she remarked.

"I have other things on my mind," said Fiona. "I'll take it up again one of these days." Fiona was lying, she had promised herself that she would never play again, not if dragged to the piano by her mother in full scale hysterics.

"Met him at a dance. He's a junior. He's going to be a lawyer..." Fiona recited these facts as if she were reading his resume out loud. "He's on the hockey team, his father is a doctor, he has three sisters, the youngest one is my age..." Everything but his height and weight, her mother thought.

"So what do you like about him?"

What a question. What doesn't she like about him? In fact, she would be hard put to describe him except in very general terms, blond hair, grey or blue eyes. "He's got kind of a nice smile," she told her mother, proceeding cautiously; she wasn't sure how hard it was going to be, her argument; her mother was unpredictable in some ways and Fiona herself, not particularly resourceful. Plans of attack, organized resistance, these were beyond her capabilities. It would be some years before she planned ahead.

Fiona managed to talk with Mark almost every day. Crouched on the little stool by the telephone table, whispering into the receiver, she actually looked as if she were fastened to the telephone. She spoke in a quivery, buttery voice, wavery, wobbly, it was as if she were almost too weak to talk, as if her only source of strength came from the speaker at the other end of the line. Everything she talked about appeared to be secret, but since the telephone calls came every day, it seemed unlikely that they only talked about secrets. Evidently, Fiona talked to him in that perfect little baby voice habitually.

"He can actually hear you when you mumble into the phone like that?"

"Oh I think he can, yes, I'm pretty sure he can," said Fiona, vague and uninformative. Vagueness was a new tactic for her who only recently had stopped meeting the adversary head on.

In fact, Fiona thought she might be pregnant. A few days later, she learned she was wrong, but at the moment, she thought it was a good possibility. The pill was still ten years in the future and Fiona had no diaphragm. She knew next to nothing about contraception or conception and neither did any of her friends at school. Highly suggestible, she had some salami sandwiches and a milkshake on the train the night before and felt sick afterwards. It could have been morning sickness, she thought.

Certainly this possibility leant steel to her argument. Fiona was now able to return her mother's hypnotic stare, in fact, she met it rather easily. Gazing steadily at her mother, she noticed the whites of her eyes; they were bloodshot, faintly yellow. "It's the sherry," she thought all of a sudden. "Oh my God," she said to herself, and inspected her mother's face so intently that it was her mother who finally broke the stare.

Fiona sat crosslegged on the floor playing solitaire. If she turned up hearts, Mark loved her, high hearts, or better yet, an ace, he loved her very much. Diamonds meant money but Fiona didn't care about money. Spades meant a fight; Fiona already knew about the fight. Her mother came in with a glass of sherry, which she drank standing up. "Fiona," she said, lowering her voice almost to a whisper, "I forbid you to go."

Fiona kept on laying out the cards, clickety click, spades and more spades, there were certainly no surprises in this hand. "It's all arranged," she said finally, eyes fixed on the cards. She moved a jack onto a queen. "I already bought the tickets."

"I don't know these people," said her mother, setting her glass down and moving out into the middle of the living room. "I don't know anything about them."

"Father is a doctor," recited Fiona. "They live five miles north of Portland. One of his sisters is my age..." She moved a red five onto a black six and turned up the ten of spades.

Her mother sat on the chair by the fireplace and said in an even softer voice, not a whisper, "I forbid you to go."

"I heard you the first time," said Fiona, and smiled knowingly at the cards.

Her mother left the room. She's gone to get more sherry, Fiona told herself. She looked down the long ell of the living room where her mother's two pianos stood against the far wall, music scores stacked on the bench next to a row of chairs where students sat awaiting their turn at the piano. Her mother played duets with the more advanced students, Fiona too of course: Clementi, Mozart, Haydn. There were portraits of composers on the wall next to the pianos, Brahms, standing beside a mountain stream, hat in hand, Beethoven, his face tilted sullenly downward, the baby-faced young Mozart with his little sister. Fiona hated these pictures, she would have liked to turn them all face against the wall. During her first months at school, she dreamed she was returned to this room for life, forced to sit straightbacked on the piano bench and play a Clementi sonatina over and over, her feet on the pedals encased in ugly shoes. She laid out a new hand of solitaire and regarded the cards grimly, they were a jumble of hearts, diamonds, and clubs with nothing to tell her. She swept them up off the floor and into a drawer and took out her history book, barricading herself from her mother.

"I want to talk to you," said her mother, lowering herself to the sofa by Fiona's chair. The vacation was coming to an end.

"Right," said Fiona, nose in her book.

"Look at me, Fiona. I want you to look at me." It was her mother's mad face, lips pinched together, the buttonhole mouth, Fiona called it. Eyes boring into Fiona's face as if into her very soul. "You do not have my permission to go," she said.

"I don't think I need it," said Fiona.

Her mother raised her voice. "What do you mean, you don't need it?"

"To get on the train," said Fiona, raising her voice in turn. "I say, I don't think I need your permission to get on the train," but as she spoke, she pictured her mother at the station, weeping, dishevelled, running after her, tugging at her bag, calling out to the porter, announcing to the world that her disobedient daughter was abandoning her.

"You live in my house, you go by my rules," said her mother, furiously.

She could run faster than her mother, thought Fiona, she could run up and down the cars, in and out of the bathrooms, up to the dining car and back, her mother would never be able to follow her. "What's that got to do with it?" she asked.

"Everything," said her mother and paced down the length of the room to her piano where she delicately pressed down three keys producing a sharp little chord. Then she came back to stand by Fiona. "You cannot disobey me," she said. She paced away and then back again. "You cannot do this," she said.

Fiona watched her. "She's going to have another fit," she told herself. After awhile she said, rather plaintively, "I don't know why you have to get so upset."

"Upset," said her mother venomously, as if the word fell so far short of the truth that it was actually insulting. She was beginning to breathe unevenly, moving back and forth across the living room and into the hall. "I'm not upset," she said finally in a broken voice. "I'm concerned. I'm very concerned. It's my job to be concerned."

"Why do you have to make such a big deal out of everything?" asked Fiona despairingly. She would never be able to tell Mark about this, or anyone else, ever, she thought to herself.

"You'll do as I say," shouted her mother. She came up to Fiona and put her hands on her shoulders, ready to shake them then and there but Fiona turned herself round and slid out of their grasp. Then she went into her room where she lay on top of her bed, fully dressed. "Jesus," she said aloud. "Oh Jesus." She lay with her eyes open staring at the ceiling.

Her mother pushed open the door and came in and stood by her bed. "Do you hear me, Fiona? Are you listening to me?"

"I'm not deaf," said Fiona.

"Don't ever, ever talk to me like that," said her mother and grunting with the effort, hit her across the face on her mouth, breaking the thin skin of her upper lip. Blood trickled into the corner of Fiona's mouth and ran down her chin. She touched her fingers to her mouth and looked at the blood, then she put her head back onto the pillow and gazed steadily upward at her mother, unblinking.

"You can't really stop a person," she said and was prepared to say more but her mother was retreating, sobbing, horrified, moving back to her room. She had no sooner left than Fiona sprang out of her bed, went to the bathroom, and checked out her mouth in the mirror. Still bleeding. She washed it off with cold water and made a compress out of a folded-up towel, then went back to bed holding the towel to her mouth. "Oh what a bitch," she whispered to herself. "What a bitch, what a bitch." Presently she began to smile as she lay there, and she kept on smiling, handkerchief still pressed against her lip, until she fell asleep fully dressed, sneakers and all.

The next morning, Fiona took a taxi to the train station. Her mother stood at the door, erect and dignified, she was her piano-teacher self again. Fiona leaned out from the taxi and waved, then sat back, ready for whatever was coming next. She was wearing a black chesterfield coat and a tiny hat, like a bonnet. She had dressed herself carefully, the swelling on her lip worried her and she had covered it with a thick layer of lipstick but it was still puffy. When she got on the train, she went into the bathroom and examined her lip but the bathroom mirror was grimy and half-crazed and she could only see her face dimly.

The train was crowded but she found a seat at the end of the car. She had started menstruating early that morning and had now almost forgotten she ever worried about getting pregnant. The war with her mother she pushed to the back of her mind as well. Instead, she gazed dreamily out of the window, watching the winter countryside, hard-frozen and bare, sliding past her window. As the train moved northward, fir trees in increasing numbers lined the tracks, dark green and edged with white, and she found the empty and uncomplicated countryside harmonious and pleasing.

Poets Don't Do That

by Don Dubuisson

ONE HOT WEDNESDAY AFTERNOON MARYJANE SAT down on a park bench and lifted her feet off the planet. Behind her lay the great expanse of Central Park filled with a bubble of voices in a variety of languages and in front of her rose scorching co-op buildings dripping with sweat from air conditioners at top speed. Her legs were swollen, she knew that, but she was now convinced she had made a mistake coming here. She put her feet down and noticed for the first time that she had on men's shoes.

As songbirds circled overhead and yellow taxis circled around the park, Maryjane moved to the end of the bench, careful not to drag her long black coat through a family of ants who now lived in a melted Eskimo Pie. Maryjane pulled her shopping cart closer and reached in to pull out a big straw bag. Not too long ago this bag hung in an open-air shop in Mexico. It then rested on the shoulder of a shopper who was still seeing the faces of starving people south of the border as she peered into glass front refrigerators in her local bodega. Now the bag held what was still precious in Maryjane's life.

"Not today," she whispered.

From within the bag came a tiny voice.

"I know I promised," Maryjane whispered, "but what if someone were to see?"

"Please...you promised," she heard the voice say.

She looked around to see if anyone was looking. People flew by in the back of cabs and there was the sound of children in full throttle. On the corner a man sat on a box beside his soda cart. His head rested on his fist. He was asleep. Now was the time. So with great care, the care a stamp collector would use to handle a one of a kind rarity, she pulled out of the bag her son, Alphonse. He was the size of a AA battery.

"One look. That's all you get," she whispered as she opened her hands and held him up.

"I don't see anyone like me, " she heard Alphonse say.

"I told you. There is no one like you." She put him in one hand so she could adjust the suit she made for him with the other hand. He stood still, motionless, as she straightened his red jacket.

"You look cold. You have to bundle up." Maryjane's mind was now back in time when she and her husband, Franklin, first came to New York. Before that they had lived in Tuscaloosa. Franklin had tried to be many things: a farmer, a house painter, even an inventor like his namesake, but he was forever sad—said he just wasn't like others. Then, out of the blue he came home announcing he was now a poet. He had, as far as she knew, never written a poem, not even a bad one. But here he was telling her they were leaving Alabama to go to New York because that's what poets do.

Maryjane could feel Alphonse moving in her hands. "Easy...easy." She blew her warm breath into her hands.

Once in New York Franklin never did anything he thought poets didn't do. Things like filling out forms, or standing in line, or speaking to the landlord about the back rent...

"Poet's don't do that."

So she had to.

But he would beg for a few dollars and come home Tuesday night with a second-hand record player and worn out records so he could twirl Maryjane around their little apartment. Or he would borrow a car to drive miles to see some tree that turned solid gold only one or two days a year. They spent a lot of time in cheap bars and graveyards and on park benches.

"Alphonse...Alphonse," for a brief scary moment Maryjane thought she had let Alphonse slip out of her hands. But she looked again and there he was...

Franklin never did write a poem. He never cashed a check. He never knew how they lived. He always said he wanted to die as a poet. Who knew what that meant? Who knew it meant having a stroke and dying nerve by nerve? At the end, Franklin was fused to the bed, unable to move a muscle.

"Look, Mommy. Look at the lady in the funny hat. She looks like it's going to snow."

Maryjane had just enough time to see the hand of a small girl leashed like a dog to a lady in pink shorts. The hand tried to take Maryjane's woolen cap. She flinched and in that single moment she dropped Alphonse, her son, somewhere in the dirt.

"Come, sweetie, leave the lady alone," a voice said above the pink shorts.

Maryjane dropped on her knees and shot her hand underneath the bench and patted the soil frantically. Tears and sweat rolled down her cheeks. She bumped her head several times.

A pair of policemen came over to offer assistance. "Need some help there?"

Maryjane pulled up into a ball. She moaned out loud.

"What's the problem, sister?"

"Nooo," Maryjane moaned. "Leave me alone."

"Hey, Bill, help me out here." One of the policemen reached to lift up Maryjane but in doing so he stepped on Alphonse.

"Nooooo," screamed Maryjane. She then bit the cop on the leg.

"Yo, lady. Call a back- up, Bill. We got a live one."

The back-up team arrived and put Maryjane screaming into the patrol car. She sobbed into her woolen cap—the same one she wore when she first came to the city.

"What d'ya think, Bill? Takes all kinds, don't it?"

"Yeah. New York. Just when you think you've seen it all."

The next morning two teens were teasing each other as they walked down the sidewalk next to the park. The boy kept telling the girl he wanted to take her out. She kept telling him to dream on. As they got near a bench the boy thought he saw a perfectly good shotgun shell. " Hey, cool," he said, but the girl ignored him.

When the boy reached down to get what he thought was free ammo, he found instead a AA battery wrapped in red construction paper and tied with white string. "Damn. Here," he said throwing the battery at the girl. She caught it and unwrapped the paper. " I hope it works." She put it into the back of her Walkman radio. At first all she could get was news—another stabbing, a rise in unemployment, suicide. Then she found some music and her body moved to the delight of the boy. Again and again he begged her to go out with him, but she pretended not to hear.

Lynol's Children

by Elana Greenfield

LYNOL LAY UNCONSCIOUS ON THE FLOOR OF THE helicopter. His heart had died and been reborn a parachute. He knew this by the feeling of reeling and lightness in his chest. How shall I ever return, he asked his heart, and he slowly descended.

On a broken stone step sat a woman. She was dressed in a long piece of silk. Her dark hair fell down her back. It was Jaffa. She was sobbing.

Oh go away, she said when she saw Lynol. Go away now. The children are not here. They are out fishing. Who are you, she asked, bored and broken hearted.

Her hair, Lynol noticed, was long, it ended at her waist. He was amazed.

It doesn't matter, said the woman. Your children are not here, she yelled. Nevertheless the boats will soon come in. Go down to the docks you will find the children there. They are precocious and they are thieves but they are also good-natured and friendly. Now don't talk to me anymore. I am glad you are handsome, she said, and put her head on her knee and went to sleep.

I am handsome, thought Lynol, but of course there's been a mistake about the children. Someone had a fever, it was me. He squatted down beside her, rather his legs buckled, so he watched her sleep. She was beautiful, she wore gray. His heart was melting in his chest.

I love you, he said. I will accept your children as my own.

Do as you like, said the woman. I am very sleepy.

He touched her hair, he continued down the sides of the city towards the docks. The water near the docks, he noticed, was a brilliant red. It was a red window, through it he saw a red whale which he knew in his heart was white. This is wonderful, he thought, and began to run. He swooned away, it seemed he moved forever back from the window, the memory receded, the impact did not happen. There was something he loved. He would let nothing hurt it. There was a woman in his arms but she turned out to be a box. No matter he thought, soon I will eat. At the end of the street he found the table, on it were onions and bread. This filled him with grief. It is unbearable, he thought. His legs were numb. I must throw water on my legs, he thought.

By the harbor he leaned against a wall. The air was strangely white. My chest collapsed, I could not think clearly, he said to the court that was made up of three whales, and a fourth to the side. They moved their foreheads and looked at him in silence. Love gone wrong, he thought.

The boats floated into the harbor. Behind them they pulled silver nets through the water, which he noticed was becoming gold. He could barely contain his excitement. His children appeared on the boats. They appeared so silently out of the mass of still moving fish at the bottom of the boats that at first he thought they were merely fins. The woman stood up high on the cliffs of the city, she turned and moved quickly towards the water. She was talking to him in her own language which he understood yet sounded strange to him. You have three faces, she was saying, naturally you are a butcher.

His children came off the boats. At first there seemed to be four or five but later it turned out there were more. They walked past him like an army, skinny and brown. They are skinny, he thought, they do not look like me. One or two of them had opaque eyes as if some metal had been melted into the eye socket, eyes used to see silver fish in gold waters. They are strange, thought Lynol, the woman is a witch. Yet his heart was swooning. Snow covered the ground. He lay wrapped in heat. On a hot plate there was water boiling. It filled the room with steam. In a state of semi-waking his body wracked by pleasure, he saw from a great distance a spot of light. It grew into a city on a hill, a city whose streets were beautiful and crooked, a city banked by gold waters, a city filled with his children who were skinny and whom he did not know.

Regret

by Garrett Kalleberg

I AM SITTING AT A DESK WEARING A PLAID SHIRT. Often I wear a white shirt. In fact I do not in general like plaid shirts yet here I am, sitting, wearing a plaid shirt. Yesterday I wore a white shirt and I sat here, at my desk, which is black. If I mention my desk it is because I am sitting at my desk tonight and I am writing this—that I am "wearing a plaid shirt." How odd. You would think that I would be wearing a white shirt, because this is what I often wear, and that, wearing a plaid shirt, I am hardly myself tonight. And yet I am. In fact if I were not wearing a plaid shirt tonight, at my desk, then I would not be myself. Because I am sitting at my desk, wearing a plaid shirt.

I am he-who-wears-a-plaid-shirt-now-and-is-writing.

I am he-who-sits-at-his-desk, -often-wearing-a-white-shirt:

I am he-who-wore-a-white-shirt-yesterday. In fact if I did not wear a white shirt yesterday I would not be myself. Because I wore a white shirt yesterday. And before that, I am trying to think, before that I was wearing a black shirt, that was the day before yesterday, though I am not proving my point very well that I often wear a white shirt, lately, because I am he-who-wears-a-plaid-shirt, -who-wore-a-white-shirt-yesterday, -who-wore-a-black-shirt-the-day-before-yesterday. And if I did not wear a black shirt, a particular black shirt, which I like, made of fine corduroy with a short brass zipper, the day before yesterday, I would not be myself. Because I wore a black shirt the day before yesterday as I walked for a while for a good several hours up and down the street in the rain. Actually, there was no rain the day before yesterday, I am confused, the day before yesterday I was sitting at my desk, thinking about writing, wearing a black shirt and, if I was not wearing a black shirt yesterday, I would not be wearing a plaid shirt tonight, while sitting at my desk, writing, and now drinking a glass of water or, alternately writing and drinking, writing and drinking as suits my fancy. Because I am he-who-sits-at-his-desk-writing, -having-worn-a-black-shirt- the - day - before - yesterday, - now- drinking - a - glass-of-water, -now-writing-etc. The question may be asked: And what if I did not wear a black shirt the day before yesterday? Or for that matter: And what if I did not wear a black shirt on July 7 1991 (I am only guessing that I did, though there is a high degree of probability that I wore a white shirt, as is often the case), and what if I did not think to go on, that day, in the still heat of a tiny room, sitting at my desk, go on writing, in my white or black shirt—

And I did go on, I must have gone on. The day may be remembered and spoken of as if it were the past. It is not too difficult even to form in one's mind the idea of such a room, a tiny white room with a black desk and books and a tiny bed place. A room, or a generic place in which one can be, as to eat sleep etc. A white room, as I say, as white is often said to be neutral, blank, empty, ready to go, ready to accept the events, come as they may. Such is the room to which I was bound, by necessity. No matter, I have since moved, I have taken my desk with me and my shirts and whatnot, the rest. I am here now, I am writing, no need to belabor the rest.

And if I were not sitting here at my desk now wearing a plaid shirt and writing, you would not be here too, sitting here, reading. This is a fact, is it not? Consider the following: if you were not here, where would you be? For the sake of argument one may say that a body (this term will do) exists, and any body that exists must exist in a particular place and a particular time. This may be said of bodies: one cannot exist and not exist. One cannot be in one place and time and likewise be in another. This should suffice. Therefore if I were not the writing-body wearing-a-plaid-shirt-body sitting-at-desk-body and so forth—if only one could grapple with the sum of all attributes!—then you would not be at all, because you *are* here, now, sitting here, reading this. No matter that you are not exactly sitting: you would not be at all, if I were wearing a white shirt, you would not be at all, if I were wearing a blue shirt, you would not be at all, if for one moment I had worn a shirt other than that which I had worn on any particular day, the above July 7 being cited as an example, no matter that I'm mistaken and here may interpretation enter in, you would not be reading you would not be thinking you would not know your left foot to be where it is you would not bear your tongue in such a way that the teeth do not bite it off you would not smell whatever it is that you might be smelling in this place where you are, in spite of yourself and here an old trick: "Don't think of elephants!" No matter, you would not have blinked your eyes in an elusive moment of boredom as you did, not willingly nor without any sign of will whatsoever, not once and then again quite a long time ago, in the past. I am being rather general, I admit. All the same you would not be at all if you were not the-one-reading-what-I-have-written-where-I-am-he-who-is-sitting-writing-wearing-a-plaid-shirt.

You may suspect this is a kind of philosopher's joke such as that where two mirrors are placed face to face in a white room and may be said to mirror each other. You may suspect this is a literary trick where writer and reader reflect one another, where writer and reader may be said to be bound, by necessity, as I am bound to write for I am at my desk I am sitting I am writing. Which gives rise to mention: and so you are bound to the situation you are in, considering, that if you were not in your situation, sitting in your own sort of way and occasionally blinking and again mildly suppressing the ludicrous thought of elephants—you would not be at all. Nor would you have been born, because your birth is only one of the things that must be said to be true about you in order for you to exist—nor would your mother have been born, even, for your mother is the-mother-who-gave-birth-to-you and you are the-one-who-blinked-again, -and-then-again, -who-now-reads-what-I-have-written. The same may be said of any one of the existents. I put it in the form of a proclamation: without this act, even, as every letter appears on the page in which I write, you would not exist, nor your mother before you, nor even your first ancestor, nor even the first cell, germ, or unit of life, or monad.

And from the point at which this thought is possible a connected order may be discerned, bound up with the very first moment as with the very present. From this very point a system of things, a connection or congruity or an agreement may be discerned, as of the order of words which brings them to bear meaning. Then when I write, "I am sitting at a table wearing a plaid shirt," it will be understood that this is a necessary abbreviation of a narrative linking all things and all times. And tomorrow, when I wear a white shirt, it will be understood to be a certain inevitability.

April 13 1994

Stance

by Jesse Murray

wherever this is, I am
wherever I am, this is

this is the point—where
    the inside and the outside meet
    an emptiness is filled

    an emptiness must be filled
    a fullness must be emptied

    a home is not home
    a place is not a place

    an emptiness is not an emptiness

    an inside is outside

wherever it is, I am
wherever I am, it is

At Dusk It Is Etc.

by Jesse Murray

How even death
transfigures darkness
as the body feels
the ultimate
weightlessness
of being and softly
softly with great
tenderness is dying
into light
uplifting
To be uplifted into
the ultimate
weightlessness of being
softly softly
this dying
into lithe
mourning
In the morning
that never comes,
may never come
that will not leave
their place
shades shadows voids
the shadows

are numberless traces
illuminating the dark

Only this morning
the light approaches
gently as if it wanted to
be near you in intensity
relation to the sight
of day returning a
blue ridge along the
edges of things a shifting
mist relinquished to
the shade

Signs of Progress

by Marin Gazzaniga

"HE SAID HE'S IN A VAN," JEANNE GRETCHUM told her daughter. "A white van."

"Hmm," was all Lora could manage. Lora Gretchum had spent years begging, demanding, cajoling—wishing her mother would go someplace to get help. Now that day was here and she couldn't think of a thing to say.

"He'll be here in an hour," her mother added.

Lora rose from the table, abandoning her tuna on iceberg lettuce. For once, she wasn't hungry in her mother's presence. "God it's freezing," she said, huddling under the old comforter on her couch.

"Your ceilings are so high—heat rises," Jeanne said. "And those windows...you need some plastic."

"I know," Lora sighed. They'd discussed ways to make Lora's apartment warmer before; it was easier to tackle than her mother's dwindling bank account, her deteriorated physical state, her hungover tremoring.

Lora watched as her mother wandered the room, drifting toward her two neatly packed duffels, stooping to rearrange the contents yet again. Earlier, they'd ventured out, Lora supporting her mother, still weak from the dt's. They shuffled from store to store picking up last-minute supplies as if they were preparing a kid for camp: a pair of tennis shoes, shampoo, sweat pants.

"I'm bringing my blue robe," Jeanne announced, holding up the garment Lora had bought a few days ago after checking her mother into the hospital. "For luck."

Lora's glimpse of the robe re-triggered the terror of watching her mother's body rebel—right there, in her apartment—of seeing it fight off sobriety the way most people's bodies fought a disease. She'd felt helpless, yet responsible, somehow, knowing it was for her benefit that her mother had tried to go cold turkey for the weekend.

Lora rose and went over to squeeze her mother's round shoulders. "This is what we all want, you know," she said, wanting to put words to what was happening, speaking for her sisters who lived far away.

"I know," her mother answered, her eyes tearing up. She pulled away and moved to the window, scanning the street below. "I wonder if it will have a big logo on the door," Jeanne joked limply. "A big cardboard sign on the roof."

"Probably," Lora joined in. "'Drunk tank.' An army-green mini-van."

Her mother made the face the Gretchums made to reprimand bad puns.

When the phone rang they flinched. Jeanne looked at her daughter, then picked up the receiver. They both knew it was for her. "Hello?"

Lora turned away at the sound of her mother's soft, artificial phone voice.

"Yes? Oh you are!" her mother looked outside. "Do you see the Greek coffee shop across from the playground? We're right above. OK. I'll be right down."

She hung up and headed for her coat. "He's here. Right

across the street. He has a car phone!" she exclaimed, as if this sign of progress meant something about her future. She struggled into her coat and scarf. "I'll say goodbye up here, so he doesn't have to watch." Jeanne held her arms out for Lora, her fingertips quavering.

Lora wrapped her mother in a hug. "Let me get my bags," her mother said, pulling away again.

"I'll get that Mom. Let me help." Lora hoisted a duffel onto her shoulder, and they headed out.

A dark-haired man in a ski parka was opening the sliding door of a white van. He took the bags from mother and daughter and tossed them in the back. Her mother thanked him repeatedly and was about to climb in when she turned back abruptly. Lora was right behind her. "Oh! There you are. Thanks. Thanks for everything. I'll call you."

Lora had a fleeting urge to shake her violently, to scream "Do you know where you're going?" Instead, she held her tightly and whispered, "I love you," enunciating each word as if it held a secret meaning for both of them. "Good luck."

Her mother turned and climbed in. The man slammed the door shut behind her. The wind was bitter and Lora had left her coat, so she dashed up the flight of stairs to her apartment and rushed to her window.

Later, when Lora would visit, her mother would tell her about all of the white van jokes people made up there. Later still, Lora would create a joke of her own—one about blue robes—after she'd bought her mother another blue robe, for another relapse. "Mom needs a new blue robe"—that would be her joke, her code phrase, for the calls to her sisters whenever it happened.

But on that first day, Lora didn't yet see her mother's struggle in terms of blue robes. In the descending dusk, she saw the white van, with her mother in the back seat and the man in the ski parka driving, get stuck at the red light at the corner. In that extra moment she was able to watch, she pictured her mother and the driver, alone for the two-hour trip upstate. What would they talk about? When the light turned green and the white van pulled away, Lora waited for a feeling of celebration. Instead, she felt something inside of her being yanked, as if the van was dragging some part of her behind it, the empty cans attached to a bumper. She watched the van for as long as she could, until she couldn't see it anymore and all she could see in the blackness was reds, greens, whites, and blues, the primary colors of the lights outside.

Axiom

by Lois Gallagher

having opted for not weeping
where a rare freedom between
departure or arrival of the same
is an accident of each

a form of muscular laxity between lake and land

(sometimes the road dips although it is true
east and west both lie
                                        as far behind as ahead)
—the anticipated result: a sentence
having eaten its way into life
is habit
shearing/through the multiplicity of things
where bodies having long since been possessed anyway
are reflections
becoming themselves only   at a remove

a parrying woman/a woman in flight
as (a woman who once butchered
herself   a closed   kissed   forgetful
mouth.   smile.)   she
used to say
                                    confidently
some feelings bite always in the same place

Healy

by Jennifer Natalya Fink

YES, YES, MRS. HEALY.

Yes, yes, Mrs. Healy, I repeat after her, the "*s*'s" sticking to my tongue. I still crave that game, crave the final "s" that grips tongue to mouth until you're so sticky with it she has to let you go.

I started speaking early, quickly, and in paragraphs. Nonetheless, the lisp saturated my speech so thoroughly that only my sister Jane could completely understand me. Even my parents couldn't quite decipher my lisping breathy prattle. Jane became my official translator. I spoke and their eyes would turn to her. Little girls grow out of these things, my parents reasoned.

Little girls lose extra syllables like baby fat, they told themselves.

In first grade at Northern Elementary School I went to see Mrs. Healy after recess at 1:30 on alternate Wednesdays in the white and green trailer house attached to the main beige building. It was just that sort of '70s rural school: cows in the front yard fenced off from the playground, trailer house attached in the back for Special Ed. I remember nothing of how I unlearned my lisp there. I do know for certain that on each alternate Wednesday, Mrs. Healy wore a different shade of sparkly blue or pink or green eyeshadow all the way up to her thin arched brows. And the stick. She never threatened or hit with it; she simply pointed it at the hard "s" words on the pages of my reader. It was too thick and long to be easily manipulated, so it would jostle all around in her hands as she'd point, its orange-varnished surface glinting with the light from the fluorescents overhead.

Yes, yes, Mrs. Healy.

Yes, yes, Mrs. Healy. Jane started the game, and "Yes, yes, Mrs. Healy" immediately became its official name. All of her games had official names. It was the summer I was obsessed with finding hidden dumbwaiters in restaurants and stripping down to my underwear in the cold empty auditoriums in the posh seaside scientific community where dad had taken us so he could do his research. Julia was always Mrs. Healy. The other researcher's kids were younger than her and boys and liked the idea of messing up Daddy's forbidden lecture hall. We were all Miss and Mister to her when she stood up there, hair pulled into a tight bun, white skin perspiring around her thin lips. With the long pointer she'd stand at the lectern and assign us our tasks. Whatever she said, you had to respond, "Yes, yes, Mrs. Healy."

I was the projectionist. My job was to follow Mrs. Healy's motions with the slide projector, projecting the hot oval light slightly below her neck so that she wouldn't be blinded. She would yell, "Lights!" and I would twirl the projector onto her thin lips as I responded with "Yes, yes, Mrs. Healy," carefully shaping each "s" with my tongue and teeth. She glowed in that oval, the metal braces on her front teeth spraying the light back over us all as we waited for our next instructions. I'd hide in the corner of the lecture hall behind the rows of blue fold-out chairs with my slide projector, excitedly pulling off my clothes, waiting for certain failure and its stick.

I liked my job pretty well, especially the part where I'd get tired and forget to say "Yes, yes, Mrs. Healy" and be called up by her in her white rage to take the stick. I'd close my eyes as the orange wood tickled the back of my throat where my tonsils had been, thinking of the blues and pinks and greens of Mrs. Healy's eyelids as she pointed out the final "s's." It never splintered but sometimes Jane would have to have a string of kids in a row take it: "Mr. Andrew! Mr. Scott! Miss Jennifer! None of you are addressing me correctly! Have you forgotton the rules?" and we would line right up, swallowing each other's saliva with the wood, not daring to spit it out in her presence.

Yes, yes, Mrs. Healy.

"Say it." Her rage was smooth, quick, absolute. The beads of sweat would gather on her part, a shallow river dividing the two halves of her brown haired skull as she'd bend over us and the piss would squirt out of my thin grey underpants. Drying off in my corner under the oval projection light I would murmur "Yes, yes, Mrs. Healy" to myself, wanting it to never end, wanting to stand beneath her forever with the saliva of the other kids dripping slowly against my throat where the "s's" now emerge clearly when bidden at the end of each word.

A Reading Problem

by Mary Gordon

A handsome man sent me a note:
"I enjoyed talking to you at the party.
It would be nice to see you once again."

I didn't answer, because that day
I was holding a beautiful book.
It was written by a man who,
in his all too brief life had answered
far too many notes suggesting dates
and places and who suffered
far too many nights of jealousy and rage.

It was a book I wanted nothing more than to read in peace.

But the handsome face of the man
who wrote the note to say
he'd like to see me
shone like the sun and blotted out the letters
of the words of the beautiful book until.
I couldn't even see them, never mind read or understand.

I find it isn't what I have the heart for.
All of it: the complications,
and the handsome face, the fair hand

keeping me from the beautiful words I long for
like good bread, like grapes in winter
like the nectar
of the gods.

Interview with Grace Paley

by E.M. Broner

GP: My older sister's piano teacher bawls her out for not practicing. So finally my sister says to her piano teacher, "I'm not practicing to play in Carnegie Hall." So then this eighty-seven-year-old teacher says to my eighty-five-year-old sister, she says, "I don't teach anybody who isn't practicing to play in Carnegie Hall." So that's a story.

EB: This interview is for the *Global City Review* issue "To Make A Long Story Short" and Linsey Abrams and I, your student and colleague , thought of you as the first person to include in this issue. Linsey said you get more into a three page story than other people get into whole novels. And she wonders why a series of simple scenes and fragments taken together become a complex whole. I think what she's asking you is how a fragment, a 'slice of life,' becomes ultimately a story of love and morality, and she wonders how you arrived at this way of writing stories.

GP: She'll wonder a long time because I'm not telling her. [Laughter] Just to begin at the end—I used to say that all of life should be in almost every sentence, let alone in every story. Now, of course, in a way that's just a thing you say; you know, it can hardly ever be done. But the term 'slice of life' gives me the willies, so to speak.

I guess it's something I never was interested in. How you do it is the mystery of our common task. You don't know exactly how you do it...but I do know that I'm not really satisfied with a story, even if it's two pages, unless it does some of that work. Every story has some work to do.

EB: I love that. Every story has some work to do.

It's clear from *The Collected Stories*, that wonderful new book, that a change has occurred. As you have aged, so have your characters. There are the earlier stories that deal with the young mother and her children, the young woman and her sexual appetites, and the child with the loud voice becoming braver and different from her parents. Do you think the stories become more philosophical as the author is striving to explain her own voice in "A Conversation With My Father" or with her place in her life as a middle-aged writer in "The Long Distance Runner"?

GP: Well, I don't know. It's funny, when I look at the whole thing, it seems like it could have happened at once. I think probably a lot of writers feel that way looking back. It seems like it's all part of a whole in some way. So I can't say that I've gotten more philosophical—But you know, in a way it's up to the critic or the reader to say that; it's really not up to me.

I know that I began not with a young woman but with a generation preceding me, which is "Goodbye and Good Luck". I mean, I began to write about someone who was in my past rather than my future. And I've always been more interested in two or three generations in a story as opposed to just one. So I guess I have more of a historical feeling than a psychological feeling.

In "A Conversation With My Father", it seems I'm being very historical. We're arguing out a historical position—and a geographical one, since he comes from another country where nothing is possible and here I am, an American girl, where much is possible. So it seems that that was the argument...but a lot of stories have arguments. You have arguments with somebody in your head. They used to say—"There should be conflict"; well, I really don't like that word at all. But it *is* dialectical. You have oppositions, some of them very mild, and some of them very serious. That's what happens in a story.

Almost all stories are conversations that you're having, not just with your characters but, maybe, with somebody else about your characters—even if you don't know it all the time. In a story called "Friends" about the death of a friend of mine, I felt I was talking to someone else about that whole event, not so much with the characters themselves.

EB: All right, well, you were just talking about the story "Friends". I was thinking about "Ruthie and Eddie" and "Zagorsky Tells". How many of your characters tell and listen? I think about you in the same way I think about Barbara Meyerhoff, for whom the tale was very important, and who taught a course on the "tale bearer." Her students would go and interview the old Jewish people in Venice, California, and come back altered. So, telling and hearing are particularly essential to the writer.

GP: Yeah, well, in the last book there are stories—one is called "Telling", the other's called "The Story Hearer" and the third one's called "Listening"—in which the teller or writer is not listening. [Laughter.] Hasn't been listening for many years.

EB: I once asked you jokingly at some event, "What comes after post-modern?" and you said, "Death." [Loud laughter.] I think about the collection being so beautifully dedicated to Sybil [Claiborne.] And I'm wondering what happens after the three books? Is there a sense of loss that one carries...that you would carry now into the next phase of writing?

GP: Well, there's something really bad about getting these three books out. I mean, what's bad is that that seems like *it*. You know, I said to them, "What do you want to that for? I'm not dead. I'll probably live yet for a few years." They said, "No, no, we want to do it. We think it's a good time to do it." So I said, "Well, all right." I mean, I have an obedient streak in my disobedient nature. But I still feel like whenever you finish a book, unless you're well into another book, which many people are, you really feel like you'll never write again. And not only that—when I never write again it's because I forgot how to write, I don't know how to write. It's always like you have to start again. You have to learn all over again. The truth is you're learning on a harder level.

EB: Oh, that's comforting. [Laughs]

GP: Well, the reason you're learning on a higher level is you know one of the most important things and this is what cheered me when I began to write my second and third stories. I knew that if the story was vital, was forced by pressure, I would carry it through to the end.

EB: Do you have a sense of subject matter or is it...where is life going to evolve inside of you?

GP: Well, it just seems to me I'm going to continue doing practically the same thing I've been doing. I can't imagine changing—I just can't imagine really doing things much differently. On the other hand, there are breaks between the books. I mean, there is the sense that I left the house so to speak and went out into other neighborhoods and into other countries. Which is what I did in one of the books. So I imagine I'll do more of that. I should think that that would be very much a part of my life and my thinking, meeting with other people, not *necessarily* family anymore.

On the other hand, I've led my life mostly in political communities and in communities of women and men and in communities—as time went on—mostly of women. So it seems that those are factors that will probably play some role.

But then also when you get older you do a lot of looking back. And that's been good. I just recently had to write a ten page—like people say "slice of life." They used to say "memoir"—but I really hate that word. I mean, we write all the time. But I did sit down and just remember six days at the Women's House of Detention—which I hadn't done before. So I'll probably be doing little things like that...mostly things that I think are useful to others.

EB: We used to have these long rides, long talks on the way to Sarah Lawrence College. We talked politics, we talked family, and we talked religion—

GP: We had a half an hour each way of pure pleasure.

EB: You said then, when we talked about writing, that the danger is in sentimentalizing your past. So how does one remember with rigor?

GP: Well, first of all, one of the biggest lies in the world is sentimentality...So once you make kind of a vow always to be truthful, then you really can fight it. It's an untruthfulness...if you hear sentimental music you're hearing a lie. So you have to just fight that lie. You know: how cute I was or how cute how everybody was. [Sarcastically] How wonderful it was—

EB: [Equally sarcastic] "How happy I was as an adolescent!" [Laughter.]

GP: But you know, I can imagine writing something which has a certain degree of sentimentality. I'm just really thinking of our car rides...Sometimes I'd ride up with Danny Kaiser, and I had very fine rides in which he'd give me lectures on Yeats — {Laughter} that he was going to give in the next two hours at school. But apart from that, he would talk about camp. He would talk about camp with the greatest pleasure, with the greatest joy. "This was wonderful, that was wonderful..."

So how would you deal with it? He could tell that story and somehow...I'm just trying to think technically now...how would you avoid the sentimentality of it...because he really did enjoy it. So what you have to show is what he thought was so great, like the greens and the yellows {camper colors} murdering each other. You know, it would be kind of fun to have someone tell that story. This is where you get into someone else's voice...how great everything was and how wonderful it was and everything like that. And yet you see that that voice is not the most dependable in the world...you see that it was these kids really about to tear each other's eyes out. You come with your own knowledge against this other person's knowledge.

EB: Nice.

GP: It's like—just an example...in "Goodbye and Good Luck", she's in love with this guy all the time but you see what a schmuck he is. And every time he opens his mouth, she thinks it's the most brilliant thing anyone ever said.

EB: Along that same line—when we were at school together —because students are all limited by their contacts in the world, I would get stories which were racist or deeply sexist, and I would quote them your remarks. You would say that the writer has to be larger than the character. The character can say something unacceptable and terrible but you must have a sense of the intelligence of the writer.

GP: Well, it's not so much the intelligence because the writer has to be more innocent—or dumber, in a lot of ways. But that's one thing. On the other hand, the character can be a shit or mushily sweet. The writer has to know what's happening in her or his world. On the other hand, we know that writers were totally vengeful bastards. We know that every book that we've ever read in nineteenth century literature is anti-Semitic.

EB: Especially if it's in English.

GP: Especially if it's in English. But we know that George Eliot wasn't [anti-Semitic.]

EB: Think of Daniel Deronda.

GP: So we learn from that that we as writers have to look at everybody in the world with some kind of compassionate truth.

EB: That's lovely.

GP: I mean, at that time brilliant people...even into our century, even into the twentieth century—were anti-Semites—

EB: All the British mystery writers, for instance.

GP: Oh yeah, all them. But also very smart people, great poets like Eliot...everybody. But you know that there were people who weren't like that. We were talking the other day about slavery and somebody was saying, "Well, the Quakers had slaves." And yes, they had slaves until at a certain point John Woolman understood that this was a terrible thing...There were others who thought of it, but he suddenly felt that as a Quaker that was not acceptable and then he spent the rest of his life travelling all over the place, visiting Quaker households in the south and the north.

We have to as much as we can look at the world with justice—let me take back the word compassion—I would say we must think in terms of justice. Justice as a way of illuminating every character so that you can see them whole.

EB: I think about George Eliot coming out of a very rigid fundamentalism and yet finding a kind of connection to Zionism, to another vision. She had left her earlier rigid world and more and more entered a world where she became peripheral to the mores of the times.

GP: —the bravest person in the world. So was George Lewes.

EB: Well, as to your work—is it still political and in a way utopian? Do you still hope for a felicitous world, an integrated city, love equally distributed? Or have you lost hope?

GP: You lose hope when you don't do anything. You lose hope because as soon as you become hopeless or despairing, then you sit down and you find all kinds of excuses not to do anything. I mean, I think the world is scary. It's much scarier and much more terrifying than I believed it would be. And certainly what happens in Europe is scary and what happens in Africa.

Look, when we first heard about Rwanda, what did we hear? We heard about a bunch of maniacs, they're all crazy, they're all killing each other, they're crazy people down there. Then, little by little, we learned what the politics of this thing is, that the government trained a militia that would act at the drop of a hat to kill the other people. They received their weaponry and instruction from the French. It was a political act and these people were trained meticulously and over a long period of time to be killers. They wanted killers.

So when you learn that...it gives us some information. It's not just that human beings are fucking maniacs, but that they were trained in this way. Like the guys who are running Haiti were trained here in the School of the Americas to be the people they are. So that gives you information and something to struggle with and against. Somebody is putting these people up to this stuff. So then you have to find out: who's the somebody? I mean, we knew at one point that in Ethiopia and in all those places that it was the Russians and the Americans. Now it may not be the Russians and the Americans necessarily, although all the guns in Somalia, I think, were Russian and American guns. So we know that people are running around the world selling arms. Even Pavel is doing that for Czechoslovakia, and when you ask them they say, "Have to make a living."

EB: I was just thinking. To go back one moment, you mentioned already about the six days in the House of Correction. Maybe for the readers you ought to elaborate on that just for a moment.

GP: Oh, sure. I was arrested for sitting down in front of a military parade or something like that. I think it must have been the May 30—but for some reason I wasn't part of the trial with the other people. I was tried with one other woman. Everybody else got one day or something like that. But the judge was so angry at me because I was already even then an older person compared to the other people...because...what was it? It was the middle sixties, so I was in my forties, early or mid-forties. And he was mad—I was a mother, I had kids at home and I was a teacher and what was wrong with me that I should do a stupid thing like that—so he gave me six days, which was very hard because it was over July 4th weekend.

And if you get time off, whatever it is, they can't let you out in the middle of a holiday because there's nobody to let you out. So there I was in jail and I was there for six days.

And it was one of the many fortunate things that have happened to me. Without a lot of my political friends around me, I just was around women who were addicts and prostitutes. I mean, that's all, that's all they were on that cellblock. Almost all of the women who were there in the building were in for stuff like that. Some were there for forgeries, check forgeries ...Every one on my block was black or was Puerto Rican. So I had the good fortune to be able to sit there and listen to these people talk and answer them and speak to them and so forth.

EB: I hear you speaking and it makes me think of your saying, "I chose to be a happy person."

GP: That was long ago. I told you that?

EB: Don't you believe that?

GP: No, no, it's not that, it's just...I remember being so unhappy—all through high school—so miserable. And at a certain point I had to either go on being miserable or I had to make a decision not to be—to refuse some of that. And I did at some point. It's not that I'm happy all the time. Things are so hard in life around you. If it's not—it's your children. I mean, you can make a decision to be a happy person but if there are problems with your children, happy-schnappy.

EB: [Laughter] So you have no further instruction on happiness?

GP: Except for your children you should be happy. [Laughter] Except for the problems. No, it relates only to the happiness of your children—If your children are miserable, forget it, you can't be happy. [Laughs] But if other people around you are miserable, you can pull yourself together somehow and be nice to them and cheer them up and make soup and everything. You know?

EB: You've been working on poems and stories with a religious them, like the story "A Midrash on Happiness" and the poem "Five Day Week"—

GP: I wrote "Five Day Week" for Esther.

EB: The first draft is framed on my wall. But both pieces, the prose piece and the poem, are religious. Is this a surprise to you, to go this way, to become biblical in cadence?

GP: No, I think I always had that because I always read—I was going to say the Bible but what I mean by that is I read the King James Version from very early in my youth. The stories of the Bible seem to have been something...in my atheistic house...I don't know, it just seems that my parents or somebody must have read them to me...because they were very close to me, always. And then as I read more I really—at some point I was writing about writing and I said that if you read nothing but the Bible, you could make literature [out of it.] So it's always been something—the sound—interesting to me.

EB: Do you remember my giving the MFA's variorium editions [of the Bible]? I passed out different translations of *B'reshit*, 'In the Beginning' at school. That was a very interesting teaching device. Four or five or six different translations of the same passage.

GP: Yes, I loved what you did.

EB: I want to talk about teaching—I taught for twenty-four years at one school [Wayne State University in Detroit and about four or five years at Sarah Lawrence] and I often used to ask myself: I taught but what did I learn? I mean, I always had a sense that literature was being written, it wasn't in the past, but there on the table in front of me, and that was very good for me. You've had a couple of decades at Sarah Lawrence or more. Were you informed by your students? Do you want to talk about the act of teaching?

GP: It was a surprising opportunity for me, first of all, to teach...because I don't have any degrees at all. So it was very good and brave of Sarah Lawrence to take me on...At the time I only had one book. I had to get over the surprise of making a living, to begin with. Which was always part-time and only half a living anyway.

EB: [Laughter] And no health insurance.

GP: I really try to think what I got out of it—but it seems like all I did was make friends for life. [Laughter.] I don't mean just the teachers like you. There are kids that are still my pals, my buddies, who are writers who are interesting to me, who became colleagues...and others who didn't but who did other things which was also something marvelous.

It gave me an opportunity to think about literature in a different way because I had to talk about it. So that was a gift which I wouldn't have had otherwise. I didn't do it in my lifetime, not having gone to school. So that was another thing...that I could think and talk with these people about it, about literature and poetry. And sometimes you'd do things you just only liked to do—like I would have this one afternoon a week for a few years where all we did was just read poetry. We would just gather, about seven people and we'd come and we'd just read poetry. I mean, what could be nicer than reading Milton at great length with people who wanted to do it...and not discussing it afterwards, which I've always hated doing.

So that's what I got out of it. And then the gift of continuous contact and relationships with youth. So as you get older, there you are, you're still talking to young people, new young people. Sometimes you know what's happening better than people who are ten years older than they are, who are out of touch with them. You're more in touch—I've talked to a lot of freshmen up at Dartmouth this year—I'm teaching them and then I met some freshmen from Sarah Lawrence. And I knew I could tell people who had graduated five or six years ago what was happening with the freshmen. I felt close. That was a gift. And there's politics again. So to talk to a bunch of freshmen who are young women who don't say, "Oh, yes, I believe in equal rights but I'm not a feminist" or some smarmy remark like that—is great. And kids who are interested from high school on in writing and in literature.

EB: Do you want to talk about the people who—writers that inform you now, whose careers you have watched with interest?

GP: I always say, "There are many writers I have been crazy about all my life and I've been a big reader, so I've read an awful lot, but these are the four books I've read in the last period of my life." Well, the four books...but then of course you want to mention a couple of other people. [Laughter.]

EB: Four and a half books.

GP: Four and a half books. Well, the two books that I read most recently were *Ceremony* by Leslie Marmon Silko and I read Michael Ondaatje's *The English Patient.* What else did I read? I read a book by a guy—Howard Norman—*The Bird Artist*—I just read it in galleys. That's really, really lovely. And

I read Annie Proulx's book...and I read it for many reasons, one of them is she lives up the road, very close.

EB: Her newest book?

GP: Her newest book is *Shipping News*, a wonderful book. Annie Proulx writes mostly about men. I don't know whether it's because she has three sons or what but she does well by the men. Actually, the man she's writing about is so fakatke ...crazy that...strange. So I'm just thinking about it.

EB: I know PEN women [from the organization of writers, PEN] the Women's Committee would say, Who should be on the panel? You seem to have the most access to the younger writers.

GP: There was a period where all I could remember was *A Weave of Women*. There was a period when people said, "Is that the only book you've read?" I know this really great book by...[pointing at Broner, then addressing the other person in the room]. I mention it again and again because she hasn't written another one [like it] and I'm very angry at her. A very early student of mine though I had nothing to do with her writing so well is Spanidou. Irini Spanidou. *God's Snake*. It's a stunning and amazing book. It's a great feminist book.

EB: That's a painful book. A couple of things: I once wrote a review in which I talked about the characters of your books. I felt that they became family to the reader. I then felt that one could almost say that you were really writing a serial novel. I presented this to you and you were—I think rightfully—irritated by this.

GP: You could say it. It's okay to me. I'm less...as I said, it seems like I wrote it all in one minute.

EB: Yes.

GP: Like life, it happens in a big bathtub. It just all happened at once. So, yeah, I think it's legitimate to say that. I mean, I would say that certainly...at one point, Farrar Strauss wanted to publish all thc Faith stories, and that would have made it really seem like that. And that's what I didn't like...because there are lots of other lives and lots of different people. I had a couple of the Irish stories, the Black stories. I didn't want that.
EB: One question I want to ask—maybe even the last—Linsey Abrams had asked a question about your splitting your time between Vermont and New York. What part would you say the city played at shaping your subject matter? Are you a city writer? I would amend it to say that actually your poetry has been largely about Vermont and about travelling. Am I wrong?
GP: Well, a lot of it is. I mean, I have a lot more poems that I wrote sitting on the porch looking at flowers, different flowers. It's true.
EB: Is it harmful for your art to be away from Eleventh Street, Sixth Avenue, P.S. 41, Washington Square?
GP: It could be, it could be. I think what happens to me is, it's good and bad. It's good for my life that I become happy in another place, that I love it and can spend...I mean, I spent the last week really—look at my nails—gardening. And I'm very deep in that life with Bob.

With *my* Bob. My friend Sybil had a Bob. We were always saying "my Bob", "your Bob" "my Bob". Vermont has its own richness. Again, having lived in New York in a neighborhood always—when I was a child in a Bronx neighborhood which was very tight and now in the Village which is a neighborhood—it seems to me that I now live in another neighborhood almost more than that I live in the country. I live in a neighborhood of different people...not many of them Jewish. It's very different in that sense...I've had to really go out to the Jewish community.

EB: I know when I'm at MacDowell Colony I light Sabbath candles. I'm the only one in my cabin so I have to mark the day for myself. But then I grew up, unlike you in the Bronx, in isolation. In the midwest in my time, it was hard to be Jewish.

GP: Well, I don't go to services in New York but I go in Vermont. At least I make holiday services and I stay in touch. And I will go to a conference...the Conference of Rural Jews, which happens in the northern part of the state. People come from Canada. They call themselves Rural Jews but a lot of them come from Montreal.

EB: Is there another last word on the short fiction form or what you got out of it or why you came to it?

GP: Well, if you want me to talk about that I will. I'd been writing poetry, that's what I'd been writing. I'd been writing poetry...and not very long poetry at that...I mean, sometimes a page, page and a half, not more. So at a certain point I decided that I had to write stories, that I wasn't able to put into poems what I needed to put into stories. I think I've learnt since then how to do it but I didn't know then. So I began to write stories and use that narrative form to talk about mostly women's lives and other stuff. So in order to begin to write prose when I decided to do it...I naturally was not going to suddenly go from writing poems to writing a novel. I did not have that long view in the habit of my head. So what I did is I wrote one short story and, as I said, I finished it. And that was very exciting. That was the most exciting thing that's happened to me in all these years—finishing one story. Then I wrote another and I finished that one. So then I felt I could write.

But by then I had gotten into the habit of completion within a year. So I can't even say I chose the short story. I began with it and it chose me in the sense that I could do it. And you only...do what you can do...I could do it, so I did it. [Laughs.]

EB: Lovely, lovely. I love the story choosing you.

City of Zinder

Niger, 1969

by Kathleen Hill

IT WAS DELISSE, ONE NIGHT DURING THE HARmattan, who talked about the fort. He stopped by for a drink, stepping through the gate with a flutter of embarrassment disguised as arrogance, raking the blond hair back from his forehead. While gusts of hot wind scattered the sand and lashed the branches of the oleander bush to the ground, we sat on the verandah and talked. Between us, the flames of the candles in their black clay candlesticks from Myrria, dipped and straightened. Before he got up to leave, he had told us about the aging parents in the house in Lille where he had grown up, the Aubusson tapestries hanging in the salon, the marble fountain choked with leaves; about his one attempt to return to France and sleep again in beds where the heavy linen sheets embroidered with the family's initials never lost their chill. He had lasted a year, he said, and returned to Zinder to marry a Hausa woman whose father called him "camel eye." With these words, his face had leapt into view; pale blue eyes, surrounded by faint eyelashes, looking out as if across a distance; the corners of the mouth twitching up in a kind of perpetual grimace that turned into an outright smile when he communicated something particularly painful.

When he first arrived seventeen years before, he told us, the French military presence dominated Zinder. From our house, all the way up to the Catholic mission, the street was lined with bistros to which maison des filles were attached. The militaires would come down from the fort on the hill to have their biftek, their glass of wine and their "fille." As he described the militaires he gestured comically, mimicking their paunched figures and hooded eyes. Then he said simply: "Ah, I have been very disappointed in the French. "Was it true, what he said, that one hundred militaires coming down from Agadez in a camel caravan would require a retinue of three hundred women, and that much of the bodily deformity that abounds in the town could be attributed to syphilis?

But it didn't begin there, he said, lowering his eyelashes. Before the French arrived, Zinder had been known to do not a bad trade in slaves. Oh yes, we could be sure it was true. His in-laws had told him so. Had we seen the occasional ostrich behind a wall, enormous black and white birds on thick legs? These were leftovers from the 1870s and '80s, when fashion in all the capitols of Europe had demanded feathers. At that time Zinder had been known for its ostrich farms. Camel caravans equipped in Kano would stop in Zinder long enough to load the feathers before setting out across the desert. At Tripoli they would be taken by ship across the Mediterranean and on to Paris, Berlin, London. And then, Bah! Fashions had changed. By the 1890s ostrich feathers were no longer wanted and the farms around Zinder had disappeared. But who did we think had worked these farms, he asked, grinning.

Zinder, during the harmattan, was filled with the sound of coughing. Behind every wall people lay shivering with fever, unable to eat. By day, the wind never stopped, a high hot wind that blew the vultures backward through the sky and filled the air with a fine filter of dust. The lines of walls cast ambiguous shadows; the yellow neem leaves that lay scattered in the sand were blown to confusion by gusts that rose abruptly and set the door and shutters banging. Every morning, three blind old women sang at the gate, faces raised to the sky, chanting the same words in one strong voice. Who would honor Allah, the source of all mercy and kindness, with a gift to his poor? Who would remember those who had risen to a day of hunger? Led by a little girl, they held tightly to the ends of three short sticks that linked them together in a chain. When Zara and Lizzy ran down to give them something, one woman raised a fist above her head to wish them a long life. Then they turned away, their shawls flapping behind them in the wind like great dark wings.

It was then, too, that waking slowly at night, we heard a solitary, plucked, music that seemed to speak of the terrible unknowability of this place where Arab, European, and the old desert races had all mingled and strained to bring forth the blind child who extended a gaping bowl at noon, as oblivious of the past as of the future. Wounded city, wounded child, wearing its history like a bruise. We had no idea of anything. But Delisse knew more. So we forgave him his exaggerated manner and the way he cut us short in the middle of a sentence. What could not be forgiven was the pain in his eyes that the lips pulled back over his bad teeth did nothing to dissemble.

Buchenwald Landscape

by William Allen

Eerie silence of the woods. One lone yellow birch,
whose seed fell after the mass graves were turned,
grows as lank and tall as the forest of firs
that shadows it. At the gates, the temperature
drops ten degrees; when the mist parts at noon,
you can be sure you see settlements below
in the towns of Ettersberg and Ramsla.
Surely then if anyone had peered at the sky
they would have known! If some old queen
or Bulgar Jew cries out and nobody hears...
how well do you know there was no sound?
Kilns pour black lather into Nephele's clouds—
droplets of something with a human smell
soak my overcoat and send me shivering home.

The Ordeal of Civility

by Edward Hirsch

Father sliced the bread on Friday nights
        his knife dripping gravy
but what knife carved and sliced the son?

There were no blessings over the meal
        as the mother flickered
behind electric light, and 150 generations

receded in shadows. He felt sickened
        by the haberdasher's hands
doling out large, bloody portions of meat

but was not allowed to leave the table.
        "Thou shalt not" the Law
commandeth. "Eat first, talk afterwards."

The boy tried to muscle through his plate
        and thought of Isaac's
gagging obedience as he was smoothed

on the altar. "We are like olives"
        the Talmud says, "only
when we are crushed can we yield

whatever is best in us." Yes, but
        the son of a butcher's
son can never forget the crushing.

Thus was a sickly vegetarian scrawled
        into the Tablets' margin
while his father devoured beef and bread

        and mice feasted on crumbs.

Sighted in the Plaza

by Barbara Selfridge

IN SAN JUAN I WAS SHOCKED TO SEE MEN WEARing moustaches. What were all those Chicanos doing in Puerto Rico, I wondered. Because back in Berkeley I knew tons of Chicanos and the men *all* wore moustaches, but out in Vega Verde, Puerto Rico, noboby did.

Vega Verde's a small town, I thought. So small it has only one traffic signal. So small that the men are all clean-shaven and the women never leave their houses in shorts.

I was in the plaza, walking my after-work route from school to the post office to home when I saw the publico arrive from San Juan and Arecibo. An American couple got out, tourists, both of them in shorts, and the entire town of Vega Verde—at least those representatives hanging out in the plaza or shopping in the stores that circle the plaza—all of Vega Verde turned its head to stare at the woman in shorts.

A woman with straight blonde hair and straight long legs.

Vega Verde was a stop the American couple hadn't expected to make, but that's the nature of públicos. Públicos are taxis, group taxis you have to share. You know them by their license plates—white instead of yellow—and it's a funny thing, but on an island there really isn't such a thing as out-of-state plates so all the license plates are either yellow for private car or white for público. Outside of San Juan, públicos are the only public transportation.

Most tourists rent cars but this couple in shorts hadn't. Had they known, though, that the público would not only wait to fill with passengers before leaving San Juan, but wait again to refill itself in every town? Had their guidebook explained that in a small town like Vega Verde they would have to choose between waiting in the público and getting out to be stared at?

You could tell the blonde knew everybody was looking at her and you could tell she was annoyed. It was a billion degrees—*surely that called for shorts and who were these assholes to make her feel so undressed?* I knew what she meant, I really did, and I thought about going up to her—wasn't *I* an American? couldn't I strike up a conversation with my countrywoman to soften her time here in my narrow-minded adopted town?—but I couldn't.

All I could do was stare.

A Bullshot in Maine

from My Haircuts

by Janet Coleman

AS I CONCENTRATE ON THE INNER MUSIC OF THIS memoir, I see how deeply I believed the maxim, "The way to a good life is a good haircut." Was it wrong? Some years before attaining a literary reputation for *Fast Feet in the Black Forest*, my opus on the German theatre before Hitler, I, Rosalie Gold, was a file clerk. Yet I went to the same brownstone beauty salon as Eunice Kennedy and her famous sister-in-law. I appreciated the woodwork and the extra thick towels, and the fact that my stylist, Mr. Donald, wore a black pin-striped suit. But I did not find it exceptionally classy to learn Nadja the waxer's disapproving opinion of crotch hair, or to watch a black maid serve tuna fish sandwiches to women in Bergdorf-Goodman snow boots. I didn't like the way I fell into a mid-Atlantic accent, or dressing up for a haircut. The widow, as Nadja the waxer called her, had a shiny glazed complexion and huge unfocussed eyes. Ushered about from room to room, dressed in a smock and chemical protective gear, and smiling vaguely, she resembled a skinned and bludgeoned ferret. It made me uneasy always to be reminded this way of the dead President.

Mr. Donald's area of expertise was straight and streaked hair, the kind that looks equally good when rain-drenched or windswept as at charity balls. My hair was dark and baby fine. I had cowlicks on either side of my forehead. On their own, my bangs would not lie flat. Huge amounts of gel and strips of scotch tape were required to set them straight. In hot or wet weather, they turned into frizzballs. With hair like mine, beach weekends at best were hell.

A friend at work invited me to spend the weekend at her husband's family cottage in Maine. At lunchtime Friday, I bought an adorable gingham bikini that was just a little too small, and had Mr. Donald style my hair. We took off after work in a VW bug and by midnight hit the coast of Maine. The cottage seemed modest. I climbed the stairs to my sparsely furnished clapboard room, unpacked my clothes, and joined my friends in the kitchen for their traditional country nightcap, a vodka martini with olives and onions and no vermouth. Breakfast at the family beach club was the plan.

In the morning, I noticed that the cottage adjoined a mansion set on sprawling acres, and that my hair was all frizzballs, matted with superhold spray. I grimly squeezed into my new bikini, scotch-taped my bangs, and wrapped a kerchief around my head.

My knowledge of the sea was formed on day trips out of Brooklyn with my parents and their friends: motorcades along the Southern State Parkway that reassembled into caravans in the parking lot at Jones Beach. Carrying old Army and Navy blankets, raggedy towels, a green and white porch umbrella, a chess set, tin pails and shovels, the Sunday papers, Noxema, and wicker hampers filled with food, the party of people in eyeglasses trudged towards the shore. The day's menu included salami sandwiches, peaches, Gruyère cheese triangles, Ritz crackers, fruit punch, and hardboiled (sometimes devilled) eggs. Call me Ishmael. I spent the day making sandcastles, collecting scallop shells, and flouncing around in the ocean with Naomi Wegweiser or Jonathan Gross, and was changed out of my swim suit under a towel, so that I would not get infantile paralysis, and on the return trip, the backseat of the car would not be coated with wet sand.

As a kid, I had also spent some time in New Jersey at the Orange Crest Swim Club. My Republican uncle owned a piece of it. I remembered the vista of pink and white candy-striped umbrellas against a wash of turquoise blue. I remembered my cousins and me standing around on rubber mats in our wet swim suits eating hot dogs and potato chips that we bought for pink plastic tokens at a booth they referred to as "the concession." I remembered my pot-bellied uncle, strutting smugly about, accepting bar mitzvah invitations and coffee cake from fawning matrons lounging around their candy-striped cabanas in corseted swim suits and silver mules.

Here in Maine, aged and well-worn driftwood was the motif. The beach umbrellas were a faded navy blue. The matrons were lanky and leathery, and dressed in colors that would blend with their dogs. The fat ones dressed more gaudily, in flowered sundresses called "Lillies" that looked like remnants of a tablecloth. At ten-forty-five, a black waiter in a starched white jacket and soft white gloves brought napkins and Cheez-its to go with the bullshots the members had signed for themselves only minutes before on the driftwood cafeteria line. The buzz of capital gains merged with the beating waves.

I took my bullshot and towel down to the shore, whipped off my kerchief and plunged into the surf. Underwater, I visualized the miraculous transformation that would result from the blend of salt water, scotch tape, and hairspray. And I managed to emerge, in my mind's eye, a water-tossed nymph. I threw myself onto the sand, and let the hot sun bake my limbs. In this reverie, I heard a woman's lockjawed voice call, "Miss—." She said it several times before I realized she was calling me. My hands flew up to cover my hair. I smiled up at her, "Yes?" "Your breast is exposed," she said.

The Gardener

by Edith Chevat

SHE HAD BEEN WORKING FOR HOURS, PULLING, digging, pushing. Everything about the garden took years of work. Her niece, Terri, had helped dig and drag the mountain laurel from the woods nearby to where the lawn met the road so as to screen the house from any one who happened to pass by. Not that any one did. It was because it was so isolated that Blanche had been attracted to this part of the state. She wanted to be able to relax, walk around in her nightgown if she wanted to, and not have to worry about neighbors.

Her friend Ruth, as a way of paying for a weekend in the country, had carted the wood chips for paths to the raspberry bed. Pete, the farmer from down the road, who did odd jobs as a way of making hard cash, strung the hammock between the two old pines, mowed the lawn and cut up downed trees for fire wood. But the strip of earth along the stone fence was hers alone, her design, her work. No one helped with that.

Years before when she had first come to this part of Vermont, when she thought she and Bud would be married forever, she had given him a patch of earth for a vegetable garden. It lay, hidden and unseen, just below the rise where the house sat, beyond the row of hemlock, fir and spruce she had carted, carried and planted. She had never gone to see it or what he had done. Then when he'd moved out, gone back to the city, she had Pete cover it with topsoil and planted with grass, as if it had never been anything but lawn. She never went there.

Now she sat back on her heels and looked at her work. Something else was needed. Not that she had a particular plan in mind; she would know when it was right, when there was nothing more to do.

She began again, digging with the trowel, pulling at roots, moving bulbs, leaving tubers, piling small stones, throwing rocks into the woods behind the stone fence. A garden was something she could count on; it could not be taken away.

She got to her feet, trowel in hand. Deliberately, she lay the trowel down. She walked into the house. In a moment she was back, carrying a green container. She removed the cover and began sprinkling the powder around the phlox, folding it in with her trowel. She moved over to the apple tree, the container cradled in her arms as if it were a bouquet. In the last ten years she had put in the tree and made room for flowering bushes and shrubs to wreath the house—lilacs for spring, climbing roses, mock orange, spirea, buttefly bushes, always shoots taken from someone else's garden, plants that had proved themselves and so would more likely survive the harsh New England winter and the rocky soil of her garden.

There were no annuals or showy flowers with their brief lives, needing special care. There were no tulips or gladioli, tea roses or dahlias. She said there were enough wild flowers in the woods for her to look at—lady slippers, Queen Anne's lace, columbine, bleeding heart. She had come from the city to find peace, to get away from noise and dirt, from people who interfered in her life. She had not thought Bud would leave her, that he would die, that she would end up alone.

The house had never held her interest. Large, comfortable, functional, it was a place for sleeping and eating, with nothing of special note about it, except that it looked out upon this particular spot, a stone wall built long ago, separating woods from field. It was because of the wall she had bought this particular piece of land. Nature could go just so far, then it would have to stop.

Boulders lay nearby as if strewn by a giant hand. Pete had assured her he could move them with his tractor, if she wished. But she wanted a tamed garden set among the stones, one that showed traces of what had once been, and her imprint upon it. But no matter what she tried—dried sheep manure, chemical fertilizer, cow dung from Pete's farm—the plants withered and died. Only the irises and yellow primroses came back, year after year. Even the wild things that flourished in the woods would not grow in this place. And when by chance something did grow by itself, borne on the air, she could not leave it alone. She poked and pulled until it, too, shriveled, died. She would rather no garden than one that did not respond to her touch.

She added more dust, then began to heel the dust in with the spade, turning it carefully. She stood up, used her feet, first just the balls, pressing and grinding the earth and dust together, then her entire foot, twisting, turning, harder, harder,

This would work. She knew it.

Here was the proper fertilizer for her garden, these the ashes of her husband.

The Muse in the Kitchen: Selma Fraiberg's Model for Clinical Research

by Kathy Atlass

NEW ORLEANS, 1960

In a setting of affluence—a large house, its many rooms furnished with antiques and crystal—lived Peter, a blind child trapped inside his body in a severe case of arrested development, irrefutable testimony to the deprivations of blindness and the impoverishment of his human bonds in the earliest years. Cared for by a series of nurses in parts of the house away from his sighted older sisters and parents—even, for a disastrous summer, institutionalized in a luxurious but inadequate facility—Peter had remained severely undifferentiated, a limp nine-year-old body literally entwined around the body of his nurse on the day Fraiberg met him. Peter's case is complex; Fraiberg's notes cover her visits to him daily, five days a week, for a period of two-and-a-half years. During the same period she also paid monthly visits to Tony, an otherwise healthy, five-month-old blind infant being raised in a tiny brown house on a dirt road on the other side of New Orleans. What emerges in final form in Fraiberg's second-to-the-last published volume, *Insights from the Blind* (1977), is our first understanding of the developmental pathways blind children take and the perils of derailment to which these children are liable, especially prior to Fraiberg's revision of earlier theories in the light of what she saw, theories "thrown out one by one, like so many boring toys over the rail of a crib."[1]

PETER'S LESSONS

What Peter[2] offered Fraiberg was a new understanding of the primary organs of perception in the blind. She came with our commonsense expectation that the ears would substitute for the eyes. Instead, on her first visit, Peter came close and fingered her. Then,

> without any change of facial expression and without any show of feeling, he began to dig his fingernails into the skin of my arm, very hard, causing me to wince with pain.

From this point on, she says, it was nearly impossible to divert him from digging his nails into her or pinching her with great intensity.

> It is impossible to describe this experience. I cannot call it sadistic in him. It was as if he did not know this was painful to me, and I really felt that on the primitive undifferentiated level on which Peter functioned he was not able to identify with the feelings of another person. This digging into me had the quality of trying to get into me, to burrow himself into me, and the pinching had the quality of...holding onto me for dear life.[3]

Staying with her subjective sensations, Fraiberg began to realize that Peter's frenzied seizing and holding of human "objects" felt like the oral-incorporative seizing and sucking of an extremely young child. His pinching fingers—were they gums, or teeth? She experienced his fierce grabbing onto her neither as erotic nor aggressive but, more simply, as a life grasping to "take her in."

Fraiberg's eventual language for this was to say that Peter's hands were in "morbid alliance" with his mouth, that he had failed to achieve hand autonomy, a crucial sign and a crucial factor in maintaining personalities on the level of nondifferentiation.

> When the mouth remains the primary organ of perception, the distinction between "inner" and "outer", "self" and "not-self" will not emerge. Perceptual experience is restricted to a narrow range of objects—those that stimulate the mouth in some preferred way. The qualities of objects cannot be known.[4]

Infants open their eyes and perceive a world of objects and people, to encounter and construct, gradually sorting the "me" and the "not me" in an elaborate process of cross-modal matching of invariants—light with sound, skin pressure with temperature and texture, inhalation with smell and taste, direction and speed with muscle activity, intensity and depth. Daniel Stern leads current theorists in giving us the brilliant studies which trace such processing.[5] However, cross-modal matching for the blind[6] child is exactly the problem: Fraiberg reasoned from her scratches and bruises how much our eyes must mediate in even the simplest of human achievements, for instance, in the normal, *adaptive* use of the hand. We speak glibly of "eye-hand" coordination, observing an infant in its early months at first staring at its own hands and then soon busy making the fingers reach and grasp, all of which spatial and skeletal mapping is awry for the child who cannot see. Peter had made his own false map by attempting a clever but doomed substitution—the mouth, the palate, the lips, the tongue—for the discriminatory fineness of his eyes. By locking into "hand-mouth" coordination, he had contracted and contained, all space inside himself, and he in it.

Fortunately Fraiberg noticed that the other thing Peter routinely did with his hands was to throw things. A year or so into his treatment he still could not appreciate the differences between the silk, velvet, and corduroy bean bags his now re-connected mother worked to sew, a dozen or more. But in an unbreakable part of the house, he could throw them, and Fraiberg began to understand that this throwing was necessary to him.

> I then changed my tactics and permitted, even encouraged, the throwing without knowing where this would lead. During these throwing sessions he was undoubtedly in better contact with me and his mother and there were evident signs of release in his personality. Yet my conscience as a child analyst troubled me very much during this period as I watched playdough, blocks, and plastic toys go flying through the air in an apparently meaningless barrage. Never before in my therapeutic career had I given permission for such aimless throwing or had the feeling that it was justified for therapeutic purposes.[7]

As Fraiberg watched Peter during many such sessions, she saw that the throwing itself became more energized and lost "its aimless quality". He was throwing hard and showing evident satisfaction when the object made contact with the floor or walls. He began to make noises and to accompany his throwing with little cries and sounds. As Peter became more active in his throwing, the biting, pinching and scratching receded. When the throwing was impeded, he regressed to his former limpness, withdrawing into autistic symptoms of the earlier "undifferentiated" times.

Fraiberg guessed that Peter was beginning to achieve some "self/it" distinctions and that sound cues were now entering to help him sketch tentative boundaries. Further, she speculated that some of the aimless throwing and dropping done by sighted infants in the last half of their first year might be part of separating the skeletal muscles from the mouth. She wondered if the throwing itself might be a bridge by which the skeletal muscles organize to "take over." The familiar sight in Western countries of the baby dropping things from her high chair occurs at six-to-eight months and coincides with the advent of the reach and collapse pattern that quickly smooths out into the crawl. Blind babies do not crawl on schedule. Peter had never crawled. In her subsequent work in the Ann Arbor Study (1963-73), Fraiberg discovered how to teach blind children to crawl but that this event is dependant on their achievement of object permanence, which comes no earlier that ten months, even in the gifted. A sighted infant will simply perceive, reach, and attain an object within its grasp as early as five months, without needing or having a stable concept of the object at all. The sighted child will also begin to extend her arms in the direction which means "pick me up" at the same time, a gesture that does not appear in blind children before eleven months or even one year of age, leaving us to wonder what must be the evanescent sense, not only of self and other but of valuation of self by other, for such children.

But all that is another story. Suffice it to say that when Piaget visited Fraiberg's Michigan lab in 1967, he tossed his beret in the air and cheered at the climax of a series of film clips illustrating the painfully protracted stages in the development of object permanence for the blind child. Such jubilation is the silver lining on the terror one imagines to be at the heart of the matter. What would it be like to be born blind? How would one's development proceed, or would it proceed at all?

Over a period of eighteen years, in first a problem-finding and then a problem-solving mode, Selma Fraiberg and her research team were able to live close enough to the realities of the blind child's dilemma to be able to provide some answers to both questions. And in a move which they did not even consider resisting, they were able positively to "interfere" with the outcomes in their research population by enhancing the blind child's chances to be read and understood by their primary caretakers as soon as their cues were decipherable by Fraiberg and her staff. Fraiberg was righting "misdevelopments" as they were occurring. As she put it, "the final stage of this research is application, which brings us back full circle. The clinician-researcher is back in the kitchen..." The research she engaged in was a kind of kitchen research, her term—"close to the source of supply, inventing with the materials on hand."

> It is plain in its design and functional. But then, science is not the practice of sacred arts. If we love science we can love it for its commonness, its everyday presence in the form of observing, questioning, examining—its everyday practice in our clinical work.[8]

Fraiberg's research returns the fruits of science to the clinic, where they should be. The frequent sterility of the classical research mode comes from those who no longer love the world they study enough to begin to think about how they might serve, or save it.

NOTES

[1] Fraiberg, 1977, p. 15.

[2] With Toni's less compromised "normal" development as a constant reference point...

[3] Fraiberg, Ibid., p.28.

[4] Fraiberg and Freedman, 1964.

[5] See without fail Daniel N. Stern, *The Interpersonal World of the Infant*, 1985.

[6] ...and the autistic...

[7] Fraiberg, 1977, p. 44.

[8] In L. Fraiberg (ed.), 1987, p.99.

That's All That I Am

by Sheldon Jaffe

POPEYE AND BLUTO ARE IN LOVE WITH OLIVE. SHE represents their ideal: long and thin and shapeless, her androgyne an unconscious predicator of the sixties.

Olive is either impotent, a patient cipher waiting for the outcome of Bluto and Popeye's battles ("I'se get to azk her out!" "No, I'ses get to azk her out!") or the all powerful tease goddess flitting and flirting on her meaningless whim, one moment enthralled by Bluto's blunt strength, then returning to the pure Popeye (albeit only after he has demonstrated his far greater spinach induced powers), for her final embrace. She retains, however, the one right of women, and, on rare occasion, she may spurn them both for the sheer maddening pleasure it gives her—the only power she has. But, this aberation from the norm, however rare, is also only temporary.

"I'll saves ya," shouts Popeye.

"Wrong, runt," says Bluto, imprisoning him in ties of steel, "I'll Save Her!" And thus they fight, the primordial battle, while Olive approaches seeming certain death, closer and closer she comes to the turning buzz saw: "I'll Save Her" "No I'll Save Her"

In the background a syrupy strings rendition of "Did You Ever See A Dream Walking" plays on while the unaware Olive goes from one increasing danger to the next. "I'll Save Her" "No I'll Save Her"...

Olive never fantasizes about Popeye or Bluto when she masturbates. She has difficulty thinking about anything other than her orgasm, which always feels weak and forced. She circles her clitoris with her right index finger and, occasionally, uses her left hand to stimulate her vagina. She'd like a dildo but is too embarrassed to buy one...doesn't even know where to look. Bluto knows, but Olive can't imagine talking to him about it. She knows that to him she is just a convenient hole. Popeye is out of the question also. He doesn't even believe in masturbation for himself—it's a sign of weakness. He won't take a cold shower when he's aroused because that's the coward's way out. He takes them only after sex. As punishment.

Olive never orgasms with Popeye. The sex is furious and powerful but all for him. "Maybe I should put some spinach on it," she says one day in frustration to her therapist. "Then he'd lick it!"

Her therapist, while not strictly Freudian, refrains from giving strong advice. He believes that Olive's problem is one of image and that she will dump Popeye once she realizes that she deserves better.

"I think it's her parents' fault," says Cathy at lunch, sipping her Ellite Spring Water. She's having the salmon, Christine the leek soup and a salad ("I've got to lose five pounds"), and Denise the Coq au Vin. Denise can eat and eat, never gaining weight, although she has far more shape than Olive.

"It's always mommy's fault," says Christine.

"Not mommy. Her parents." Cathy breaks off a piece of bread and puts it in her mouth without butter. "I mean, my

God, what kind of parents name their children Castor and Olive."

"Especially," adds Denise sardonically, "when the last name's 'Oyl'." Cathy and Olive went to college together, but Olive never completed her degree. Cathy later got her MBA from Yale, where woman were supported, and is doing very well at a small investment firm. She introduced Olive to Denise, her lawyer, and Christine, who analyzes the Media Industry for Moody's. "Some positive imaging - that's what she needs." It never worked, of course.

"Olive," said Christine at their first lunch, "that outfit." She said it like it required a response, although Olive had no idea what she was talking about.

"Outfit..."

"I mean," Christine continued, "I suppose it might work for certain occasions, but surely you don't go to the office like that? I mean, this 'Country Girl' look may have its place..." her voice trailed off. Olive looked down at herself and tried to understand what Christine was talking about. Her long blue-black skirt was perfectly ironed, the roll-top socks in place, just peeking over her well-shined army boots. She always dressed like this. What was the problem? Should the red sweater be buttoned? She wished enlightenment, but the conversation had already shifted.

Every lunch she attended was the same. The others would end up discussing The Street, their diets, the pros and cons of personal trainers, and appropriate day-care options. Olive always found herself feeling like the last one choosen up for the team and then stuck out in deep left field.

"How come," Denise asked, "she's always got his nephews?"

"And that baby," adds Christine, "where did he come from?"

Cathy shrugs. "God only knows, because I sure don't." She scans the dessert menu. "Sometimes I think even Olive doesn't know." They share a chocolate mousse.

Olive doesn't feel close to Cathy anymore. She hasn't since she left school. She feels vaguely inferior, and not just because of Cathy's material success. Well, she consoles herself, at least my life has been interesting: Cowgirl, Policewoman, Conga-Dancer, South Sea Island Goddess...but nothing ever worked out.

She feels closest to Wimpy. He seems positively asexual, a fat priest in mufti, a true neuter, having successfully sublimated his repressed homosexuality through oral gratification. Hamburgers keep him happy.

Olive likes to talk to Wimpy, because he's always sympathetic, listens carefully to what she says, and, she knows, never judges her. But she's cautious with him. She knows he'll never escape his programing—she knows he'll always be Popeye's friend.

Olive's not really sure about Popeye. She thinks its some sort of socio-biological need. After all he has muscles like:

A Mule Kicking
A Battery of Battleship Cannons
A Keg of TNT
A Train Speeding
Even an Atom Bomb (copyrights past 1945)

She knows this because she's seen the films. Purchased a

collection in the children's section of her video store for $3.95. She runs the muscle sequences in slow motion.

It couldn't be his Durante nose, his mumbling gravel voice, his ferocious temper, his mis-shapen limbs (and not the right limb either, she thinks). It's got to be my genes, is all Olive can tell herself. It's got to be.

In the end, Olive always comes back to Popeye. In the end, she embraces him in the last reel, a passionate kiss, and a fade out. This is always the finish for Olive, no matter what. "Kismet," she says, and laughs without mirth.

"What's that, my sweet petunia?" asks Popeye.

"Nothing," says Olive. "Never mind."

That Leopard at Altitude

by Paul West

THE MARVELOUS THING IS THAT IT'S PAINLESS. They come in and take you in the prime of life, drawn by the smell of rotting ocelot. How they came to know such things I never found out. Off you go growling a bit and pawing the high mountain air as if to pat it into submission or friendship. Up there your muscles tend to become rigid and they make snapping, clicking sounds almost like distant gunfire, but right up under your nose. They see you from high up then come in close with the big greasy web that keeps you from leaping away. Then they bring out the Urdsweft-Brittaner, recent model, and change you into a willing subject at the same time as stroking you to soothe you trout like or salmon, except of course you are all spotted fur and claws.

While all this is going on the big birds plane above you hoping for a nibble, a clean sweep even. They have no hope, though, not with the two plug-uglies knelt beside you nor visibly breathing, on the snow beside them the square wallet of ruby ice with needles, the little light bulbs, the snappable vials, and the tiny silver seeds they feed you while working on you. If it was the same with the big bulls and the smaller warthogs, I never knew. Down below, far away, over three miles down I could see the tent and the blurred cross of the Puss Moth two-seater. We outsee even the eagles. It was I that was smelling rather then the smell coming up. After a few moments they lugged me across to the hatch, not so much a doorway as a mucous cleft into the main body of the archangel as I call it. When they want you they get you. I am the leopard. Down there was the rotting man with his woman and his friend with the plane and the whimpering hyena that was not really a hyena but a cruising destiny.

Bundled gently into the main cavern of the interior I at first thought I was somehow under the top of the mountain, in an ice ark, but it was a whole world wide and unbelievably white; quite definitely a manufactured interior, more like an empty palace than like the square top of Kilimanjaro. There was a pink sifting blizzard in there from time to time like fish scales coming down. They breathed them. They were neither leopards nor Harrys, neither vultures nor hyenas, nor even Helens or boys. They were regular guys, though, leopards at 19,710 feet don't know from guys, I guess. After the first meeting, it was all right, it was never again the hell in the mind it was at first. They brought me in, see, left me there and spun away. I never came from Kilimanjaro but from a writer's mind or another planet, a different mountain, one of the nameless ones.

And then they came and got me again once I was dried enough and frozen enough, dead as dead can be but able to be Urdsweft-Brittanered in the interior. Even when I was dead, I could always see them coming when they came or tell they were close, reclaiming their experimental animal.

They wanted a man, they said, like the hunter with gangrene way below but nobody would ever come this high tempted by the tall tale of a leopard and wondering what it was seeking at that altitude.

I was bait, see.

They couldn't understand why nobody had come up to look at me. Where they came from they climb mountains in droves to go see the frozen leopards. It is a national pastime, they having no literature, no leopard-lore, no fancy stories about the House of God.

"Molo, letti dui whiskey-soda!" they call, parroting something that has come to them from far away. It does for them as prayer, these Urdswefters as I call them. They don't have much at all and they have to go enormous distances for it, asleep and frozen, which is why I seem to be up here so long in between planting and retrieval. If ever someone, a hunter with gangrene or two bicycle policemen or any low-life form with a wide snout like a hyena, comes up and claims me having figured out what I was or am doing up here, the Urdswefters will close up shop, give up the ghost, flash away homeward in their archangel. What a hell of a breath it has when it whisks away like a wheel on fire. They will never come back until their distant receptors tell them I have been replaced on the western summit as a tribute to, and icon of, the rank mystery that envelops us all and the dark mountains ahead.

Upcoming Issues

NEXT ISSUE: SPRING 1995 (filled), Crime and Punishment explores what is forbidden by ourselves, our families, by our society; and the myriad ways in which we seek and inflict punishment. Featured in this issue are an interview with Kim Wozencraft and Richard Stratton; stories by Lucy Rosenthal, Charles Glazier and Joan K. Peters; Poetry by Rebecca Seiferle and Richard McCann; and oral histories by gay police officers.

FUTURE ISSUES: To Sleep Perchance to Dream, on dreams, fantasy, psychoanalysis, insomnia, (un)consciousness, altered states, art; and Totem and Taboo, on culture and the individual, urban legends, primitive being, primal languages, representation, the unspeakable, types of social and psychological organization, taboos of all kinds, assorted totems.

Contributors

WILLIAM ALLEN is the author of a book of poems, *The Man and the Moon.* Other poems have recently appeared in *Callaloo, Poetry East, Chelsea,* and *The American Voice.* He teaches literature at Cooper Union and at The School of Visual Arts, both in New York City.

KATHY ATLASS focuses on Infant Mental Health in her work at the University of Chicago Committee on Human Development.

SARI BRONER'S work appears in the collection *Everything is Real Except the Obvious* (Em Press) and in the quarterly, *Rooms.* She has recently completed a manuscript, *Leave to Become.*

EDITH CHEVAT has been published in *Sojourner, Other Voices,* the anthology, *The One You Call Sister,* and *Home Planet News,* amongst others. She has completed two novels. "The Gardener" is part of an ongoing series called *Portraits.*

JANET COLEMAN is the author of *The Compass: The Improvisational Theatre That Revolutionized American Comedy* and (with Al Young) *Mingus/Mingus: Two Memoirs.* She is the voice of Emily Ann Andrews on *Poisoned Arts,* the two-minute comedy series airing twice daily on WBAI-FM Pacifica in New York City.

LOUISE CORBETT has been published in the *Yankee Magazine* and the *Southern Review.* She presently divides her time between Key West, Florida and La Roq Sur Seze, Provence.

DONALD DUBUISSON was raised in the South but now lives and works in the Northeast. He is currently teaching in New York City and pursuing his doctorate in American literature at St. John's University.

CORNELIUS EADY is the author of five books of poetry: *Kartunes, Victims of the Latest Dance Craze, The Gathering of My Name,* and the forthcoming *You Don't Miss Your Water,* and *The Autobiography of a*

Jukebox. He is the recipient of many awards, including a John Simon Guggenheim Fellowship in Poetry. He is currently Associate Professor of English and Director of Creative Writing and The Poetry Center at SUNY Stony Brook.

JOE ELLIOT runs a reading series at Biblio's Bookstore and Cafe an Manhattan. He is the author of *You Gotta Go In It's the Big Game* and *Poems To Be Centered On Much Larger Sheets Of Paper.*

JENNIFER NATALYA FINK is a fiction writer, performance/installation artist and PhD candidate at New York University. Her work has appeared in *Women & Performance, New Art Examiner* and *Windy City Times.*

LOIS GALLAGHER lives and writes in Brooklyn and is currently working on a performance piece using the structure of DNA as its model and object. She has no pets.

KAREN GARTHE'S work has appeared in *New American Writing, The Brooklyn Review, The Exquisite Corpse,* and *Painted Bride Quarterly.* She lives in New York.

MARIN GAZZANIGA is completing her first collection of short stories, *What I Want.*

MARY GORDON is the author of novels, *Final Payment, The Company of Women, Angels, The Other Side,* and *The Rest of Life.* This year she received the Lila Acheson Wallace/Reader's Digest Award. She is also a Guggenheim fellow.

ELANA GREENFIELD'S plays have been presented by many New York venues. Her short fiction has most recently appeared in *On Meadway* and *Noir Mechanics.* She is Director of Artistic Programs at New Dramatists Inc. and is currently writing a radio play for WNYC's *The Radio Stage.*

KATHLEEN HILL has published fiction in *Hudson Review, Prairie Schooner,* and other literary journals. She has received grants from the National Endowment for the Arts and from the New York